The Shanghai Maths Project

For the English National Curriculum

Teacher's Guide 1A

Teacher's Guide Series Editor: Amanda Simpson

Practice Books Series Editor: Professor Lianghuo Fan

Authors: Laura Clarke, Linda Glithro, Cherri Moseley and Paul Wrangles

William Collins' dream of knowledge for all began with the publication of his first book in 1819.

A self-educated mill worker, he not only enriched millions of lives, but also founded a flourishing publishing house. Today, staying true to this spirit, Collins books are packed with inspiration, innovation and practical expertise. They place you at the centre of a world of possibility and give you exactly what you need to explore it.

Collins. Freedom to teach.

Published by Collins
An imprint of HarperCollins*Publishers*
The News Building
1 London Bridge Street
London
SE1 9GF

© HarperCollins*Publishers* Limited 2017

Browse the complete Collins catalogue at
www.collins.co.uk

10 9 8 7 6 5 4 3

978-0-00-819719-3

The authors assert their moral rights to be identified as the authors of this work.

Teacher's Guide Series Editor: Amanda Simpson

Practice Books Series Editor: Professor Lianghuo Fan

Authors: Laura Clarke, Linda Glithro, Cherri Moseley and Paul Wrangles

All rights reserved. No part of this publication may be reproduced, stored in a retrieval system, or transmitted in any form by any means, electronic, mechanical, photocopying, recording or otherwise, without the prior written permission of the Publisher or a licence permitting restricted copying in the United Kingdom issues by the Copyright Licensing Agency Ltd., Barnard's Inn, 86 Fetter Lane, London, EC4A 1EN.

British Library Cataloguing in Publication Data

A catalogue record for this publication is available from the British Library.

Publishing Manager: Fiona McGlade
In-house Editor: Nina Smith
In-house Editorial Assistant: August Stevens
Project Manager: Emily Hooton
Copy Editor: Catherine Dakin
Proofreaders: Dawn Booth, Niamh O'Carroll
Cover design: Kevin Robbins and East China Normal University Press Ltd.
Internal design: 2Hoots Publishing Services Ltd
Typesetting: 2Hoots Publishing Services Ltd
Illustrations: QBS
Production: Rachel Weaver

Printed and bound by CPI Group (UK) Ltd, Croydon, CR0 4YY

Photo acknowledgements
The publishers wish to thank the following for permission to reproduce photographs. Every effort has been made to trace copyright holders and to obtain their permission for the use of copyright materials. The publishers will gladly receive any information enabling them to rectify any error or omission at the first opportunity.

(t = top, c = centre, b = bottom, r = right, l = left)

p223 first row pencil studiovin/Shutterstock, dog cynoclub/Shutterstock, car Rawpixel.com/Shutterstock; second row sheep Eric Isselee/Shutterstock, pencil case Artem Shadrin/Shutterstock; third row Eric Isselee/Shutterstock; fourth row kitten Elya Vatel/Shutterstock, pencil case, Artem Shadrin/Shutterstock, brushes Garsya/Shutterstock.

Contents

The Shanghai Maths Project: an overview   iv

Chapter 1   Numbers up to 10   1
Unit 1.1   Let's begin   2
Unit 1.2   Let's sort (1)   7
Unit 1.3   Let's sort (2)   11
Unit 1.4   Let's count (1)   15
Unit 1.5   Let's count (2)   19
Unit 1.6   Let's count (3)   23
Unit 1.7   Let's count (4)   27
Unit 1.8   Let's count (5)   32
Unit 1.9   Let's count (6)   36
Unit 1.10   Counting and ordering numbers (1)   41
Unit 1.11   Counting and ordering numbers (2)   45
Unit 1.12   Let's compare (1)   50
Unit 1.13   Let's compare (2)   55
Unit 1.14   The number line   60

Chapter 2   Addition and subtraction within 10   65
Unit 2.1   Number bonds   67
Unit 2.2   Addition (1)   71
Unit 2.3   Addition (2)   76
Unit 2.4   Addition (3)   80
Unit 2.5   Let's talk and calculate (I)   85
Unit 2.6   Subtraction (1)   90
Unit 2.7   Subtraction (2)   95
Unit 2.8   Subtraction (3)   101
Unit 2.9   Let's talk and calculate (II)   106
Unit 2.10   Addition and subtraction   112
Unit 2.11   Addition and subtraction using a number line   119
Unit 2.12   Games of number 10   126
Unit 2.13   Adding three numbers   133
Unit 2.14   Subtracting three numbers   139
Unit 2.15   Mixed addition and subtraction   146

Chapter 3   Numbers up to 20 and their addition and subtraction   153
Unit 3.1   Numbers 11–20   155
Unit 3.2   Tens and ones   160
Unit 3.3   Ordering numbers up to 20   165
Unit 3.4   Addition and subtraction (I)   170
Unit 3.5   Addition and subtraction (II) (1)   176
Unit 3.6   Addition and Subtraction (II) (2)   182
Unit 3.7   Addition and subtraction (II) (3)   188
Unit 3.8   Addition and subtraction (II) (4)   193
Unit 3.9   Addition and subtraction (II) (5)   199
Unit 3.10   Let's talk and calculate (III)   206
Unit 3.11   Adding on and taking away   212
Unit 3.12   Number walls   218

Resources   222

Answers   226

The Shanghai Maths Project: an overview

The Shanghai Maths Project is a collaboration between Collins and East China Normal University Press Ltd., adapting their bestselling maths programme, 'One Lesson, One Exercise', for England, using an expert team of authors and reviewers. This carefully crafted programme has been continually reviewed in China over the last 24 years, meaning that the materials have been tried and tested by teachers and children alike. Some new material has been written for The Shanghai Maths Project but the structure of the original resource has been preserved and as much original material as possible has been retained.

The Shanghai Maths Project is a programme from Shanghai for Years 1–11. Teaching for mastery is at the heart of the entire programme, which, through the guidance and support found in the Teacher's Guides and Practice Books, provides complete coverage of the curriculum objectives for England. Teachers are well supported to deliver a high-quality curriculum using the best teaching methods; pupils are enabled to learn mathematics with understanding and the ability to apply knowledge fluently and flexibly in order to solve problems.

The programme consists of five components: Teacher's Guides (two per year), Practice Books (two per year), Shanghai Learning Book, Homework Guide and Collins Connect digital package.

In this guide, information and support for all teachers of primary maths is set out, unit by unit, so they are able to teach The Shanghai Maths Project coherently and confidently, and with appropriate progression through the whole mathematics curriculum.

The Shanghai Maths Project: an overview

Practice Books

The Practice Books are designed to serve as both teaching and learning resources. With graded arithmetic exercises, plus varied practice of key concepts and summative assessments for each year, each Practice Book offers intelligent practice and consolidation to promote deep learning and develop higher order thinking.

There are two Practice Books for each year group: A and B. Pupils should have ownership of their copies of the Practice Books so they can engage with relevant exercises every day, integrated with preparatory whole-class and small group teaching, recording their answers in the books.

The Practice Books contain:
- chapters made up of units, containing small steps of progression, with practice at each stage
- a test at the end of each chapter
- an end-of-year test in Practice Book B.

Each unit in the Practice Books consists of two sections: 'Basic questions' and 'Challenge and extension questions'.

We suggest that the 'Basic questions' be used for all pupils. Many of them, directly or sometimes with a little modification, can be used as starting questions, for motivation or introduction or as examples for clear explanation. They can also be used as in-class exercise questions – most likely for reinforcement and formative assessment, but also for pupils' further exploration. Almost all questions can be given for individual or peer work, especially when used as in-class exercise questions. Some are also suitable for group work or whole-class discussion.

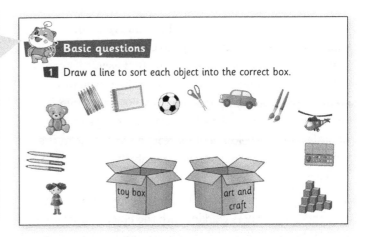

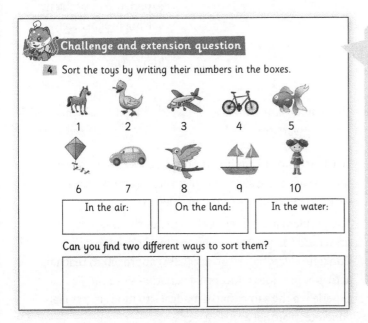

All pupils should be given the opportunity to solve some of the 'Challenge and extension questions', which are good for building confidence, but they should not always be required to solve all of them. A general suggestion is that most pupils try about 40–60 per cent of the 'Challenge and extension questions'.

Unit tests sometimes include questions that relate to content in the 'Challenge and extension questions'. This is clearly shown in the diagnostic assessment grids provided in the Teacher's Guides. Teachers should make their own judgements about how to use this information since not all pupils will have attempted the 'Challenge and extension question'.

The Shanghai Maths Project: an overview

Teacher's Guides

Theory underpinning the Teacher's Guides

The Teacher's Guides contain everything teachers need in order to provide the highest quality teaching in all areas of mathematics, in line with the English National Curriculum. Core mathematics topics are developed with deep understanding in every year group. Some areas are not visited every year, though curriculum coverage is in line with Key Stage statutory requirements, as set out in the National curriculum in England: mathematics programmes of study (updated 2014).

There are two Teacher's Guides for each year group: one for the first part of the year (Teacher's Guide 1A) and the other for the second (Teacher's Guide 1B).

The Shanghai Maths Project is different from other maths schemes that are available in that there is no book called a 'textbook'. Lessons are a mixture of teacher-led, peer and independent work. The Teacher's Guides set out subject knowledge that teachers might need, as well as guidance on pedagogical issues – the best ways to organise activities, to ask questions, to increase difficulty in small steps. Most importantly, the Teacher's Guides contain, threaded throughout the whole book, a strong element of professional development for teachers, focusing on the way mathematics concepts can be enabled to develop and connect with each other.

The Shanghai Maths Project Teacher's Guides are a complete reference for teachers working with the Practice Books. Each unit in the Practice Book for each year group is set out in the corresponding Teacher's Guide over a number of pages.

Most units will need to be taught over more than one lesson – some might need three lessons. In the Practice Books, units contain a great deal of learning, densely packed into a few questions. If pupils are to be able to tackle and succeed with the Practice Book questions, they need to have been guided to learn new mathematics and to connect it to their existing knowledge.

This can only be achieved when teachers are able to break down the conceptual learning that is needed and to provide relevant and high quality teaching. The Teacher's Guides show teachers how to build up pupils' knowledge and experience so they learn with understanding in small steps. This way, learning is secure, robust and not reliant on memorisation.

The small steps that are necessary must be in line with what international research tells us about conceptual growth and development. The Shanghai Maths Project embodies that knowledge about conceptual development and about teaching for mastery of mathematics concepts and skills. The way that difficulty is varied, and the same ideas are presented in different contexts, is based on the notion of 'teaching with variation'. 'Variation' in Chinese mathematics carries particular meaning as it has emerged from a great deal of research in the area of 'variation theory'. Variation theory is based on the view that, 'When a particular aspect varies while all other aspects of the phenomenon are kept invariant, the learner will experience variation in the varying aspect and will discern that aspect. For example, when a child is shown three balls of the same size, shape, and material, but each of a different color: red, green and yellow, then it is very likely that the child's attention will be drawn to the color of the balls because it is the only aspect that varies.' (Bowden and Marton 1998, cited in Pang & Ling 2012)

In summary, two types of variation are necessary, each with a different function; both are necessary for the development of conceptual understanding.

Variation

Conceptual
Function – this variation provides pupils with multiple experiences from different perspectives.
↑
'multi-dimensional variation'

Procedural
Function – this variation helps learners:
- aquire knowledge step by step
- develop pupils' experience in problem solving progressively
- form well-structured knowledge.
↑
'developmental variation'

Teachers who are aiming to provide conceptual variation should vary the way the problem is presented without varying the structure of the problem itself.

The problem itself doesn't change but the way it is presented (or represented) does. Incorporation of a Concrete–Pictorial–Abstract (CPA) approach to teaching activities provides conceptual variation since pupils experience the same mathematical situations in parallel concrete, pictorial and abstract ways.

CPA is integrated in the Teacher's Guides so teachers are providing questions and experiences that incorporate appropriate conceptual variation.

Procedural variation is the process of:
- forming concepts logically and/or chronologically (i.e. scaffolding, transforming)
- arriving at solutions to problems
- forming knowledge structures (generalising across contexts).

In the Practice Book there are numerous examples of procedural variation in which pupils gradually build up knowledge, step by step; often they are exposed to patterns that teachers should guide them to perceive and explore.

It is this embedded variation that means that when The Shanghai Maths Project is at the heart of mathematics teaching throughout the school, teachers can be confident that the curriculum is of the highest order and it will be delivered by teachers who are informed and confident about how to support pupils to develop strong, connected concepts.

Teaching for mastery

There is no single definition of mathematics mastery. The term 'mastery' is used in conjunction with various aspects of education – to describe goals, attainment levels or a type of teaching. In teaching in Shanghai, mastery of concepts is characterised as 'thorough understanding' and is one of the aims of maths teaching in Shanghai.

Thorough understanding is evident in what pupils do and say. A concept can be seen to have been mastered when a pupil:
- is able to interpret and construct multiple representations of aspects of that concept
- can communicate relevant ideas and reason clearly about that concept using appropriate mathematical language
- can solve problems using the knowledge learned in familiar and new situations, collaboratively and independently.

Within The Shanghai Maths Project, mastery is a goal, achievable through high-quality teaching and learning experiences that include opportunities to explore, articulate thinking, conjecture, practise, clarify, apply and integrate new understandings piece by piece. Learning is carefully structured throughout and across the programme, with Teacher's Guides and Practice Books interwoven – chapter by chapter, unit by unit, question by question.

Since so much conceptual learning is to be achieved with each of the questions in any Practice Book unit, teachers are provided with guidance for each question, breaking down the development that will occur and how they should facilitate this – suggestions for teachers' questions, problems for pupils, activities and resources are clearly set out in an appropriate sequence.

In this way, teaching and learning are unified and consolidated. Coherence within and across components of the programme is an important aspect of The Shanghai Maths Project, in which Practice Books and Teacher's Guides, when used together, form a strong, effective teaching programme.

Promoting pupil engagement

The digital package on Collins Connect contains a variety of resources for concept development, problem solving and practice, provided in different ways. Images can be projected and shared with the class from the Image Bank. Other resources, for pupils to work with directly, are provided as photocopiable resource sheets at the back of the Teacher's Guides, and on Collins Connect. These might be practical activities, games, puzzles or investigations, or are sometimes more straightforward practice exercises. Teachers are signposted to these as 'Resources' in the Unit guidance.

Coverage of the curriculum is comprehensive, coherent and consolidated. Ideas are developed meaningfully, through intelligent practice, incorporating skilful questioning that exposes mathematical structures and connections.

Shanghai Year 1 Learning Book

Shanghai Year 1 Learning Books are for pupils to use. They are concise, colourful references that set out all the key ideas taught in the year, using images and explanations pupils will be familiar with from their lessons. Ideally, the books will be available to pupils during their maths lessons and at other times during the school day so they can access them easily if they need support for thinking about maths. The books are set out to correspond with each chapter* as it is taught and provide all the key images and vocabulary pupils will need in order to think things through independently or with a partner, resolving issues for themselves as much as possible. The Year 1 Learning Book might sometimes be taken home and shared with parents: this enables pupils, parents and teachers to form positive relationships around maths teaching that is of great benefit to children's learning.

---

* Note that because Chapter 5 in Year 1 is a Consolidation and Enhancement Chapter, there is no Chapter 5 in the Year 1 Learning Book.

The Shanghai Maths Project: an overview

How to use the Teacher's Guides

Teaching

Units taught in the first half of Year 1:

Contents

Chapter 1 Numbers up to 10
- 1.1 Let's begin ... 1
- 1.2 Let's sort (1) ... 3
- 1.3 Let's sort (2) ... 5
- 1.4 Let's count (1) ... 7
- 1.5 Let's count (2) ... 9
- 1.6 Let's count (3) ... 12
- 1.7 Let's count (4) ... 14
- 1.8 Let's count (5) ... 16
- 1.9 Let's count (6) ... 18
- 1.10 Counting and ordering numbers (1) ... 21
- 1.11 Counting and ordering numbers (2) ... 24
- 1.12 Let's compare (1) ... 26
- 1.13 Let's compare (2) ... 29
- 1.14 The number line ... 31
- Chapter 1 test ... 34

Chapter 2 Addition and subtraction within 10
- 2.1 Number bonds ... 39
- 2.2 Addition (1) ... 41
- 2.3 Addition (2) ... 43
- 2.4 Addition (3) ... 46
- 2.5 Let's talk and calculate (I) ... 49
- 2.6 Subtraction (1) ... 52
- 2.7 Subtraction (2) ... 55
- 2.8 Subtraction (3) ... 57
- 2.9 Let's talk and calculate (II) ... 59
- 2.10 Addition and subtraction ... 61
- 2.11 Addition and subtraction using a number line ... 64
- 2.12 Games of number 10 ... 66
- 2.13 Adding three numbers ... 68
- 2.14 Subtracting three numbers ... 70
- 2.15 Mixed addition and subtraction ... 72
- Chapter 2 test ... 75

Chapter 3 Numbers up to 20 and their addition and subtraction
- 3.1 Numbers 11–20 ... 78
- 3.2 Tens and ones ... 80
- 3.3 Ordering numbers up to 20 ... 82
- 3.4 Addition and subtraction (I) ... 85
- 3.5 Addition and subtraction (II) (1) ... 88
- 3.6 Addition and subtraction (II) (2) ... 91
- 3.7 Addition and subtraction (II) (3) ... 94
- 3.8 Addition and subtraction (II) (4) ... 97
- 3.9 Addition and subtraction (II) (5) ... 100
- 3.10 Let's talk and calculate (III) ... 103
- 3.11 Adding on and taking away ... 106
- 3.12 Number walls ... 110
- Chapter 3 test ... 112

Teacher's Guide 1A sets out, for each chapter and unit in Practice Book 1A, a number of things that teachers will need to know if their teaching is to be effective and their pupils are to achieve mastery of the mathematics contained in the Practice Book.

Each chapter begins with a chapter overview that summarises, in a table, how Practice Book questions and classroom activities suggested in the Teacher's Guide relate to National Curriculum statutory requirements.

Chapter overview

Area of mathematics	National Curriculum Statutory requirements for Key Stage 1	Shanghai Maths Project reference
Number – number and place value	Year 1 Programme of study: Pupils should be taught to:	
	- count to and across 100, forwards and backwards, beginning with 0 or 1, or from any given number	Year 1, Unit 1.1, 1.4, 1.5, 1.6, 1.7, 1.8, 1.9, 1.10, 1.11, 1.12, 1.13, 1.14
	- count, read and write numbers to 100 in numerals; count in multiples of twos, fives and tens	Year 1, Unit 1.1, 1.4, 1.5, 1.6, 1.7, 1.8, 1.9, 1.10, 1.11, 1.12, 1.13, 1.14
	- given a number, identify one more and one less	Year 1, Unit 1.12, 1.13, 1.14
	- identify and represent numbers using objects and pictorial representations including the number line, and use the language of: equal to, more than, less than (fewer), most, least	Year 1, Unit 1.1, 1.2, 1.3, 1.4, 1.5, 1.6, 1.7, 1.8, 1.9, 1.10, 1.11, 1.12, 1.13, 1.14
	- read and write numbers from 1 to 20 in numerals and words.	Year 1, Unit 1.5, 1.11
	Year 2 Programme of study: Pupils should be taught to:	
	- compare and order numbers from 0 up to 100; use <, > and = signs	Year 2, Unit 1.13, 1.14

The Shanghai Maths Project: an overview

It is important to note that the National Curriculum requirements are statutory at the end of each Key Stage and that The Shanghai Maths Project does fulfil (at least) those end of Key Stage requirements. However, some aspects are not covered in the same year group as they are in the National Curriculum Programme of Study – for example, end of Key Stage 1 requirements for 'Money' are achieved in Year 2 and 'Money' is not taught again in Year 2.

All units will need to be taught over 1–3 lessons. Teachers must use their judgement as to when pupils are ready to move on to new learning within each unit – it is a principle of teaching for mastery that pupils are given opportunities to grasp the learning that is intended before moving to the next variation of the concept or to the next unit.

All units begin with a unit overview, which has four sections:

Conceptual context – a short section summarising the conceptual learning that will be brought about through Practice Book questions and related activities. Links with previous learning and future learning will be noted in this section.

Conceptual context

Prior to counting sets of objects, learning the numbers to 10 and comparing them, there are several skills that need to be developed. These include being able to process information presented visually and link it to known facts about objects, to understand how abstract tokens can be used to represent objects, and compare sets of different objects through subitising or counting.

At this stage, pupils need hands-on experiences – being able to handle objects as they count them adds an important physical and concrete dimension to counting. Number is an abstract concept and pupils need to experience it and see it represented in many different ways. This will help them to form meaningful number concepts that will form vital foundations for their mathematical thinking. Although the questions in the Practice Book provide necessarily pictorial representations, the experiences that you provide in your teaching should focus where possible on the 'concrete' aspect of the Concrete Pictorial Abstract approach. Pupils should be provided with opportunities to count, match and sort physical objects, both everyday objects and abstract tokens (for example, plastic counters).

It is important that, through these activities and questions, pupils have the opportunities to learn that:

a) objects do not always have to be counted – often it is possible to see how many there are without counting them individually (subitising)
b) each object that is counted is only given one number name (one-to-one correspondence)
c) numbers occur in a fixed order – the same order every time
d) it does not matter in which order objects are counted or how they are arranged (conservation of number)
e) the last number said is the total number of objects counted.

Learning pupils will have achieved at the end of the unit

- Pupils will have practised generating a subtraction number sentence from a story (Q1)
- Pupils will have revisited recording part–whole relationships in abstract representations such as bar models (Q1)
- Pupils will have used concrete and pictorial bar models to reinforce understanding of subtraction using addition bonds (Q1, Q2, Q3)
- Pupils will have consolidated using a variety of representations for recording subtraction, including pictures (Q1), bar models (Q1, Q2, Q3), mapping format (Q2), number sentences (Q1, Q2, Q3)
- Pupils will have linked the mapping format with the bar model (Q2)
- Pupils will have revisited and reinforced the correct language for subtraction (Q2, Q3)
- Pupils will have revisited recording addition and subtraction in a number sentence (Q3)

This list indicates how skills and concepts will have formed and developed during work on particular questions within this unit.

These are resources useful for the lesson, including photocopiable resources supplied in the Teacher's Guide. (Those listed are the ones needed for 'Basic questions' – not for 'Challenge and extension questions'.) A blank spinner is included as a generic resource along with the specific resources required for units.

This is a list of vocabulary necessary for teachers and pupils to use in the lesson.

Resources

small world toys, e.g. domestic animals, wild animals, sea creatures, vehicles; sets of natural objects, e.g. shells, leaves, fruits, vegetables; beads; buttons; coloured counters; coins; sets of objects made from different materials, e.g. plastic/metal/wood; 2-D/3-D shapes, interlocking cubes, hoops or sorting rings
Resource 1.1.2 'Can you see…?'

Vocabulary

sort, sets, group, similar, same, different, criteria, characteristic, odd one out

The Shanghai Maths Project: an overview

After the unit overview, the Teacher's Guide goes on to describe how teachers might introduce and develop necessary, relevant ideas and how to integrate them with questions in the Practice Book unit. For each question in the Practice Book, teaching is set out under the following headings:

> **What learning will pupils have achieved at the conclusion of Question X?**
>
> This list responds to the following questions: Why is this question here? How does this question help pupils' existing concepts to grow? What is happening in this unit to help pupils prepare for a new concept about …? This list of bullet points will give teachers insight into the rationale for the activities and exercises and will help them to hone their pedagogy and questioning.

> **What learning will pupils have achieved at the conclusion of Question 1?**
> - Pupils will have practised processing information presented visually as pictures and connecting it with concepts that they hold about everyday objects.
> - Pupils will have practised talking aloud in full sentences to describe characteristics, similarities, differences and connections related to everyday objects.

> **Activities for whole-class instruction**
> - Pupils have practised asking what is the same and what is different about groups of objects. The learning for this question requires them to find one object that does not fit a group.
> - Prepare sets of five or six concrete objects (or pictures) that are similar and add one that is different. Choose scenarios that are appropriate for your pupils. Possible ideas are:
>   - 4–5 farm animals and one tractor
>   - 4–5 vegetables and one fruit
>   - 4–5 boats and one shark
>   - 4–5 sets of shoes and one pair of socks
>   - 4–5 fish and one crab
>   - 4–5 identical birds flying and one perched
> - Encourage the pupils to work in pairs to identify the group and recognise the one that does not belong through discussion.
> - Choose a pupil to express this verbally, for example: *This is a group of (farm) animals. The one that does not belong is the tractor because it is not a (farm) animal.*
>   Repeat what the pupil has said and introduce the term 'odd one out'. Say: *This is a group of (farm) animals. The odd one out is the tractor because it is not a (farm) animal.*
> - Challenge pairs of pupils to use small world resources to make their own group of objects that includes an 'odd one out'. Invite pairs to identify and explain the odd one out.
> - Allow pupils to work through Question 3 in the Practice Book recognising the group and identifying the object that does not belong to the group. Listen to them explaining their reasoning.

Activities for whole-class instruction

This is the largest section within each unit. For each question in the Practice Book, suggestions are set out for questions and activities that support pupils to form and develop concepts and deepen understanding. Suggestions are described in some detail and activities are carefully sequenced to enable coherent progression. Procedural fluency and conceptual learning are both valued and developed in tandem and in line with the Practice Book questions. Teachers are prompted to draw pupils' attention to connections and to guide them to perceive links for themselves so mathematical relationships and richly connected concepts are understood and can be applied.

The Concrete–Pictorial–Abstract (CPA) approach underpins suggestions for activities, particularly those intended to provide conceptual variation (varying the way the problem is presented without varying the structure of the problem itself). This contributes to conceptual variation by giving pupils opportunities to experience concepts in multiple representations – the concrete, pictorial and the abstract. Pupils learn well when they are able to engage with ideas in a practical, concrete way and then go on to represent those ideas as pictures or diagrams, and ultimately as symbols. It is important, however, that a CPA approach is not understood as a one-way journey from concrete to abstract and that pupils do not need to work with concrete materials in practical ways if they can cope with abstract representations – this is a fallacy. Pupils of all ages do need to work with all kinds of representations since it is 'translating' between the concrete, pictorial and abstract that will deepen understanding, by rehearsing the links between them and strengthening conceptual connections. It is these connections that provide pupils with the capacity to solve problems, even in unfamiliar contexts.

In this section, the reasons underlying certain questions and activities are explained, so teachers learn the ways in which pupils' concepts need to develop and how to improve and refine their questioning and provision.

Usually, for each question, the focus will at first be on whole-class and partner work to introduce and develop ideas and understanding relevant to the question. Once the necessary learning has been achieved and practised, pupils will complete the Practice Book question, when it will be further reinforced and developed.

The Shanghai Maths Project: an overview

Same-day intervention

Pupils who have not been able to achieve the learning that was intended must be identified straight away so teachers can try to identify the barriers to their learning and help pupils to build their understanding in another way. (This is a principle of teaching for mastery.) In the Teacher's Guide, suggestions for teaching this group are included for each unit. Ideally, this intervention will take place on the same day as the original teaching. The intervention activity always provides a different experience from that of the main lesson – often the activity itself is different; sometimes the changes are to the approach and the explanations that enable pupils to access a similar activity.

> **Same-day intervention**
> - Provide worksheets to practise drawing numerals that use arrows to show the correct direction required for each stroke. Check which numerals individual pupils find difficult and provide focused support.
> - Write numbers on white paper using a pink or blue highlighter pen. Give pupils a yellow highlighter and ask them to trace the numbers carefully. If they trace accurately, the colour will change from pink to orange, or from blue to green.
> - Use PE small apparatus, for example, beanbags, quoits, or spot markers, for pupils to play counting games. Place the apparatus a sensible distance away. Give pairs of pupils a hoop in which to place items and then challenge them to run and collect a number of objects, for example, six beanbags. Choose a pupil to set a new challenge.

> **Same-day enrichment**
> - Split pupils into pairs. One pupil should make a group of everyday objects (all the same) to represent a number less than 10.
> - Their partner should then use plastic counters to make a group to match that number.
> - Encourage pupils to explain how they know that the groups are the same.

Same-day enrichment

For pupils who do manage to achieve all the planned learning, additional activities are described. These are intended to enrich and extend the learning of the unit. This activity is often carried out by most of the class while others are engaged with the intervention activity.

Lessons might also have some of the following elements:

Information point

Inserted at points where it feels important to point something out along the way.

> (i) This question allows pupils to identify links between objects according to their characteristics. Although these characteristics are not number-based, perceiving characteristics – and therefore similarities and differences – is a vital skill for mathematics. Pupils who get into good habits and are practised at noticing similarities and differences will be able to see connections more readily and be able to solve problems in future. Seeing similarities generally is also a pre-requisite for being able to say 'I have matched these two groups together because they both show 4' – that is, to match quantities that are represented in different ways.

All say ...

Phrases and sentences to be spoken aloud by pupils in unison and repeated on multiple occasions whenever opportunities present themselves during, within and outside of the maths lesson.

> How many more are needed to make the sticks the same. All say the sentence together.

> Look out for... ... left-handed pupils may need additional support in forming numerals, as they cannot easily see what they are writing. They may find it easier to write an 8 in reverse. Similarly, pupils with dyspraxia will find this challenging. Ensure that such pupils are allowed to demonstrate their true cognitive abilities and are not penalised for immature motor control. These pupils should be receiving support to help with handwriting and as their handwriting skills improve, so will their ability to write clear numerals.

Look out for ...

Common errors that pupils make and misconceptions that are often evident in a particular aspect of maths. Do not try to prevent these but recognise them where they occur and take opportunities to raise them in discussion in sensitive ways so pupils can align their conceptual understanding in more appropriate ways.

Within the guidance there are many prompts for teachers to ask pupils to explain their thinking or their answers. The language that pupils use when responding to questions in class is an important aspect of teaching with The Shanghai Maths Project. Pupils should be expected to use full sentences, including correct mathematical terms and language, to clarify the reasoning underpinning their solutions. This articulation of pupils' thinking is a valuable step in developing concepts, and opportunities should be taken wherever possible to encourage pupils to use full sentences when talking about their maths.

Ideas for resources and activities are for guidance; teachers might have better ideas and resources available. The principle guiding elements for each question should be 'What learning will pupils have achieved at the conclusion of Question X?' and the 'Information points'. If teachers can substitute their own questions and tasks and still achieve these learning objectives they should not feel concerned about diverging from the suggestions here.

The Shanghai Maths Project: an overview

Planning

The Teacher's Guides and Practice Books for Year 1 are split into two volumes, 1A and 1B, one for each part of the year.
- Teacher's Guide 1A and Practice Book 1A cover Chapters 1–3.
- Teacher's Guide 1B and Practice Book 1B cover Chapters 4–8.

Each unit in the Practice Book will need 1–3 lessons for effective teaching and learning of the conceptual content in that unit. Teachers will judge precisely how to plan the teaching year but, as a general guide, they should aim to complete Chapters 1–3 in the autumn term, Chapters 3–5 in the spring term and Chapters 5–8 in the summer term.

The recommended teaching sequence is as set out in the Practice Books.

Statutory requirements of the National Curriculum in England 2013 (updated 2014) are fully met, and often exceeded, by the programme contained in The Shanghai Maths Project. It should be noted that some curriculum objectives are not covered in the same year group as they are in the National Curriculum Programme of Study – however, since it is end of Key Stage requirements that are statutory, schools following The Shanghai Maths Project are meeting legal curriculum requirements.

A chapter overview at the beginning of each chapter shows, in a table, how Practice Book questions and classroom activities suggested in the Teacher's Guide relate to National Curriculum statutory requirements.

Level of detail

Within each unit, a series of whole-class activities is listed, linked to each question. Within these are questions for pupils that will:
- structure and support pupils' learning, and
- aid teachers' assessments during the lesson.

Questions and questioning

Within the guidance for each question are sequences of questions that teachers should ask pupils. Embedded within these is the procedural variation that will help pupils to make connections across their knowledge and experience and support them to 'bridge' to the next level of complexity in the concept being learned.

In preparing for each lesson, teachers will find that, by reading the guidance thoroughly, they will learn for themselves how these sequences of questions very gradually expose more of the maths to be learned, how small those steps of progression need to be, and how carefully crafted the sequence must be. With experience, teachers will find they need to refer to the pupils' questions in the guidance less, as they learn more about how maths concepts need to be nurtured and as they become skilled at 'designing' their own series of questions.

Is it necessary to do everything suggested in the Teacher's Guide?

Activities are described in some detail so teachers understand how to build up the level of challenge and how to vary the contexts and representations used appropriately. These two aspects of teaching mathematics are often called 'intelligent practice'. If pupils are to learn concepts so they are long-lasting and provide learners with the capacity to apply their learning fluently and flexibly in order to solve problems, it is these two aspects of maths teaching that must be achieved to a high standard. The guidance contained in this Teacher's Guide is sufficiently detailed to support teachers to do this.

Teachers who are already expert practitioners in teaching for mastery might use the Teacher's Guide in a different way from those who feel they need more support. The unit overview provides a summary of the concepts and skills learned when pupils work through the activities set out in the guidance and integrated with the Practice Book. Expert mastery teachers might, therefore, select from the activities described and supplement with others from their own resources, confident in their own 'intelligent practice'.

Assessing

Ongoing assessment, during lessons, will need to inform judgements about which pupils need further support. Of course, prompt marking will also inform these decisions, but this should not be the only basis for daily assessments – teachers will learn a lot about what pupils understand through skilful questioning and observation during lessons.

At the end of each chapter, a chapter test will revisit the content of the units within that chapter. Attainment in the text can be mapped to particular questions and units so teachers can diagnose particular needs for individuals and groups. Analysis of results from chapter tests will also reveal questions or units that caused difficulties for a large proportion of the class, indicating that more time is needed on that question/unit when it is next taught.

The Shanghai Maths Project: an overview

Shanghai Year 1 Learning Book

As referenced on page vii, The Shanghai Maths Project Year 1 Learning Book is a pupil textbook containing the Year 1 maths facts and full pictorial glossary to enable children to master the Year 1 maths programmes of study for England. It sits alongside the Practice Books to be used as a reference book in class or at home.

Maths facts correspond to the chapters in the Practice Books for ease of use.

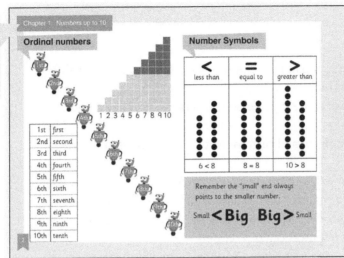

Key models and images are provided for each mathematical concept.

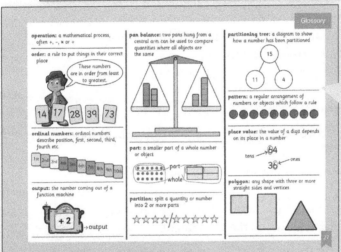

A visual glossary defines the key mathematical vocabulary children need to master.

Homework Guides

The Shanghai Maths Project Homework Guide 1 is a photocopiable master book for the teacher. There is one book per year, containing a homework sheet for every unit, directly related to the maths being covered in the Practice Book unit. There is a 'Learning Together' activity on each page that includes an idea for practical maths the parent or guardian can do with the child.

Homework is directly related to the maths being covered in class.

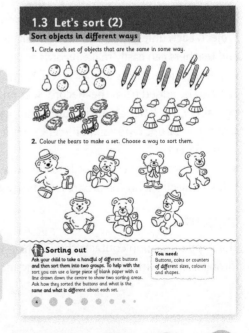

 An idea for practical maths the parent or guardian can do with the child

xiii

The Shanghai Maths Project: an overview

Collins Connect

Collins Connect is the home for all the digital teaching resources provided by The Shanghai Maths Project.

The Collins Connect pack for The Shanghai Maths Project consists of four sections: Teach, Resources, Record, Support.

Teach

The Teach section contains all the content from the Teacher's Guides and Homework Guides, organised by chapter and unit.

- The entire book can be accessed at the top level so teachers can search and find objectives or key words easily.
- Chapters and units can be re-ordered and customised to match individual teachers' planning.
- Chapters and units can be marked as complete by the teacher.
- All the teaching resources for a chapter are grouped together and easy to locate.
- Each unit has its own page from which the contents of the Teacher's Guide, Homework Guide and any accompanying resources can be accessed.
- Teachers can record teacher judgements against National Curriculum attainment targets for individual pupils or the whole class with the record-keeping tool.
- Units from the Teacher's Guide and Homework Guide are provided in PDF and Word versions so teachers can edit and customise the contents.
- Any accompanying resources can be displayed or downloaded from the same page.

Resources

The Resources section contains 35 interactive whiteboard tools and an image bank for front-of-class display.

- The 35 maths tools cover all topics, and can be customised and used flexibly by teachers as part of their lessons.
- The image bank contains the images from the Teacher's Guide, which can support pupils' learning. They can be enlarged and shown on the whiteboard.

Record

The Record section is the home of the record-keeping tool for The Shanghai Maths Project. Each unit is linked to attainment targets in the National Curriculum for England, and teachers can easily make records and judgements for individual pupils, groups of pupils or whole classes using the tool from the 'Teach' section. Records and comments can also be added from the 'Record' section, and reports generated by class, by pupil, by domain or by National Curriculum attainment target.

- View and print reports in different formats for sharing with teachers, senior leaders and parents.
- Delve deeper into the records to check on the progress of individual pupils.
- Instantly check on the progress of the class in each domain.

Support

The Support section contains the Teacher's Guide introduction in PDF and Word formats, along with CPD advice and guidance.

Chapter 1
Numbers up to 10

Chapter overview

Area of mathematics	National Curriculum statutory requirements for Key Stage 1	Shanghai Maths Project reference
Number – number and place value	Year 1 Programme of study: Pupils should be taught to: ■ count to and across 100, forwards and backwards, beginning with 0 or 1, or from any given number	Year 1, Units 1.1, 1.4, 1.5, 1.6, 1.7, 1.8, 1.9, 1.10, 1.11, 1.12, 1.13, 1.14
	■ count, read and write numbers to 100 in numerals; count in multiples of twos, fives and tens	Year 1, Units 1.1, 1.4, 1.5, 1.6, 1.7, 1.8, 1.9, 1.10, 1.11, 1.12, 1.13, 1.14
	■ given a number, identify one more and one less	Year 1, Units 1.12, 1.13, 1.14
	■ identify and represent numbers using objects and pictorial representations including the number line, and use the language of: equal to, more than, less than (fewer), most, least	Year 1, Units 1.1, 1.2, 1.3, 1.4, 1.5, 1.6, 1.7, 1.8, 1.9, 1.10, 1.11, 1.12, 1.13, 1.14
	■ read and write numbers from 1 to 20 in numerals and words.	Year 1, Units 1.5, 1.11
	Year 2 Programme of study: Pupils should be taught to: ■ compare and order numbers from 0 up to 100; use <, > and = signs.	Year 2, Units 1.13, 1.14

Chapter 1 Numbers up to 10

Unit 1.1
Let's begin

Conceptual context

Prior to counting sets of objects, learning the numbers to 10 and comparing them, there are several skills that need to be developed. These include being able to process information presented visually and link it to known facts about objects, to understand how abstract tokens can be used to represent objects, and compare sets of different objects through subitising or counting.

At this stage, pupils need hands-on experiences – being able to handle objects as they count them adds an important physical and concrete dimension to counting. Number is an abstract concept and pupils need to experience it and see it represented in many different ways. This will help them to develop meaningful number concepts that will form vital foundations for their mathematical thinking. Although the questions in the Practice Book provide necessarily pictorial representations, the experiences that you provide in your teaching should focus where possible on the 'concrete' aspect of the Concrete Pictorial Abstract approach. Pupils should be provided with opportunities to count, match and sort physical objects, both everyday objects and abstract tokens (for example plastic counters).

It is important that, through these activities and questions, pupils have the opportunities to learn that:

a) objects do not always have to be counted – often it is possible to see how many there are without counting them individually (subitising)
b) each object that is counted is given only one number name (one-to-one correspondence)
c) numbers occur in a fixed order – the same order every time
d) it does not matter in which order objects are counted or how they are arranged (conservation of number)
e) the last number said is the total number of objects counted.

Learning pupils will have achieved at the end of the unit

- Pupils will have practised processing information presented visually as pictures and connecting it with concepts that they hold about everyday objects (Q1)
- Pupils will have practised talking aloud in full sentences to describe characteristics, similarities, differences and connections related to everyday objects (Q1, Q2)
- Pupils will have developed their understanding of the use of abstract tokens to represent objects (Q2)
- Understanding of one-to-one correspondence in counting will have been revised (Q2)
- Pupils will have been encouraged to subitise and not to feel expected to count if they don't need to (Q2)
- Relevant vocabulary will have become more secure and developed (Q3)
- Pupils will have had many opportunities to count to 10 and compare numbers to 10 (Q3)
- Pupils' concepts of 'more than', 'less than' and 'same as' will be developing (Q3)

Resources

everyday objects for matching (e.g. plastic knife and fork, pen and notebook and so on); mini whiteboard; counters; interlocking cubes; objects other than counters for concrete representations (see individual questions)

Vocabulary

count, how many … ? there is/are …
number, one, two, three, four, five, six, seven, eight, nine, ten, more, less, fewer, same, compare, similar, different, match, pair, pattern

Chapter 1 Numbers up to 10

Unit 1.1 Practice Book 1A, pages 1–2

Question 1

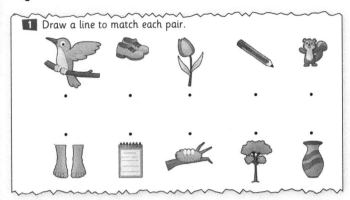

What learning will pupils have achieved at the conclusion of Question 1?

- Pupils will have practised processing information presented visually as pictures and connecting it with concepts that they hold about everyday objects.
- Pupils will have practised talking aloud in full sentences to describe characteristics, similarities, differences and connections related to everyday objects.

Activities for whole-class instruction

- Prior to the lesson, prepare a selection of ten everyday objects. These objects should consist of five pairs of objects that can be matched together, for example a pen and a notebook, an empty crisp packet and a bin liner, a shoe and a sock, a plastic knife and a plastic fork, and a tennis ball and a marble.
- Try to cover a range of reasons for each pairing – some belong together because they are used together (the knife and fork), some belong together because one object belongs inside the other (crisp packet belongs in the bin), others are different versions of the same object (the tennis ball and marble are both shaped like a ball, just different sizes) and so on.
- Separate these objects into two groups so that one object from each pair is placed in a row at the front of the class and the second object from each pair is placed in a bag (the 'mystery' bag). Explain that inside the mystery bag are some objects and that each one will make a pair with one of the objects at the front.
- Ask: *What is a pair? How many mystery objects do you think are in the bag? Why?*
- Point to one of the objects at the front. Ask: *What object do you think might match with this to make a pair? Why?*

- Choose a pupil to come to the front and pick an object from the mystery bag.
- Ask the class to suggest which of the objects it should be matched with. Encourage pupils to state their reasons for this. Ask: *Why do you think these two objects belong as a pair? How are they similar? How are they different? Does anyone have a different answer?*
- Repeat for the remaining mystery objects.
- Discuss with pupils the characteristics they chose to compare. Ensure that differences are discussed as well as similarities.
- Work through Question 1 in the Practice Book, matching images that belong together.

(i) This question allows pupils to identify links between objects according to their characteristics. Although these characteristics are not number-based, perceiving characteristics – and therefore similarities and differences – is a vital skill for mathematics. Pupils who get into good habits and are practised at noticing similarities and differences will be able to see connections more readily and be able to solve problems in future. Seeing similarities generally is also a prerequisite for being able to say 'I have matched these two groups together because they both show 4' – that is, to match quantities that are represented in different ways.

Same-day intervention

- Provide pupils with a set of objects that can be matched using the same criteria (for example objects that are used in conjunction with each other: pen and paper, glass and plastic straw, birthday card and envelope) or those that form familiar pairs.
- Allow pupils to see each set of objects immediately without the 'mystery' aspect and to discuss how they might match them.

Same-day enrichment

- Ask pupils to devise their own mystery bag activity. Encourage them to explain why each pair of objects belongs together.

Chapter 1 Numbers up to 10

Unit 1.1 Practice Book 1A, pages 1–2

Question 2

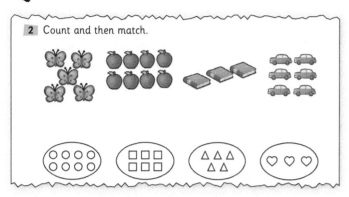

What learning will pupils have achieved at the conclusion of Question 2?

- Pupils will have developed their understanding of the use of abstract tokens to represent objects.
- Understanding of one-to-one correspondence in counting will have been revised.
- Pupils will have been encouraged to subitise and not to feel expected to count if they don't need to.
- Pupils will have practised talking aloud in full sentences to describe reasons for linking images.

Activities for whole-class instruction

- Remind pupils of the matching activity carried out in Question 1. Ask: *What were some of the reasons you had for matching the objects together?*
- Explain that objects (or groups of objects) can be matched together because they show the same number.
- Ask: *Who can show me 4 by … ?* Encourage pupils to represent 4 in different ways (for example: *Who can show me 4 by putting books in a pile? … by lining up four chairs? … on their fingers? … by clapping?* and so on). Show how pupils can match two of these representations and say: *These make a pair because they both show the same number. They both show 4.*
- Draw three small circles on the board. Ask: *Is this the same or different? Why? Is it more than 4 or less than 4? Did you need to count each circle one by one to know how many there were or did any of you 'see' the number straight away?*
- Discuss how, with small numbers, pupils may be able to see a number and won't need to count each one. Demonstrate this by holding up two books and asking: *How many?* Pupils will not need to count *one, two* to be able to tell that there are two.

(i) Subitising is the ability to instantly identify a number of objects without the need for counting individually. In the above activity, it is the ability to be able to see straight away that three circles have been drawn without counting each one (*one, two, three*).

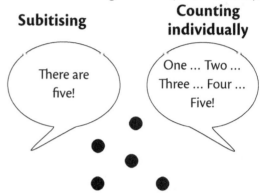

At this stage, pupils will usually display this skill when recognising groups of up to around seven objects. Recognition may occur in randomly arranged groups (pupils who recognise 3, however the objects are arranged) or by spotting familiar patterns (pupils who recognise 5 when the objects are arranged as on a dice, with four corners and one in the centre).

Subitising is an important skill to note in pupils and to develop. It saves time when counting small numbers and is the basis for more complex counting skills when dealing with larger numbers (pupils who can recognise a group in a set are more likely to begin counting from that number). Pupils who have the number difficulties related to dyscalculia are often those who find subitising more challenging. Providing opportunities at this early stage for pupils to practise subitising skills is therefore essential.

It is also important, when working with pupils, to encourage them to use these number recognition skills when they are asked *How many are there?* and not just count from one because they think it is expected of them.

- The aim of the following activity is to allow pupils the opportunity to count and match groups of both everyday objects and more abstract tokens.
- Provide groups of objects for pupils to practise counting. Ask: *How many are there? How do you know?* Move the objects so that there is more space between them. Ask: *Are there more objects now? Why?*
- Prepare five upturned plastic cups, with each one housing a number of plastic cubes underneath. The number of cubes under each cup should be different and should be less than ten.

Chapter 1 Numbers up to 10

Unit 1.1 Practice Book 1A, pages 1–2

- Show pupils a group of everyday objects. This should represent a number less than 10 (equivalent to one of the values of cubes under the plastic cups). For example, this could be a group of six tennis balls.
- Explain to pupils: *Underneath these cups are different numbers of cubes. One of them is the same number as the number of tennis balls. I want you to match the cubes with the tennis balls.*
- Ask a pupil to choose a cup and reveal the number of cubes underneath. Ask: *Do these match? Is the number of cubes the same/more/less? How did you know the number?* Encourage those pupils who 'saw' the number without counting and remind the class that they do not have to count every time. Model the counting process and remind pupils that the last number they say is the total.
- Ask: *How can we check whether they are the same quantity or different?* (For example, by placing one of each cube next to each object to show one-to-one correspondence.)
- If the numbers are equal, say: *These groups are the same. They both show _____.* If the numbers are different, choose further plastic cups and repeat the activity until pupils find the matching quantity.
- Encourage pupils to describe how they were able to match the groups together.
- Repeat the activity with a new set of everyday objects representing a different number.
- Work through Question 2 in the Practice Book, matching two groups that are equal to each other.

Same-day intervention

- Provide pupils with a structure for counting objects to ensure one-to-one correspondence and that pupils count correctly. For small objects, use a ten-frame or egg box to place each object in a different segment (in order) before counting them. For larger objects, place ten chairs in a line. Ask pupils to place each object on a different chair (in order) before counting them all.
- Use this method to begin to compare quantities. Ask: *Are there more or less teddy bears on the chairs than cubes? Can you make the same number of cubes as teddy bears?*

Same-day enrichment

- Split pupils into pairs. One pupil should make a group of everyday objects (all the same) to represent a number less than 10.
- Their partner should then use plastic counters to make a group to match that number.
- Encourage pupils to explain how they know that the groups are the same.

Question 3

3 Count the objects in the picture and then write the correct numbers in the boxes.

What learning will pupils have achieved at the conclusion of Question 3?

- Relevant vocabulary will have become more secure and developed.
- Pupils will have had many opportunities to count to 10 and compare numbers to 10.
- Pupils' concepts of 'more than', 'less than' and 'same as' will be developing.

Activities for whole-class instruction

- Without mentioning the number, choose ten pupils to sit at the front of the class. Pretend that they are going on a

Chapter 1 Numbers up to 10

Unit 1.1 Practice Book 1A, pages 1–2

class picnic at the park (or similar) and explain that it is important that they have enough food to eat when they are there.

- Ask: *How many children are going on the picnic?*
- Show the class five cards, each with an image of an apple (or similar). Ask: *How many apples are there?* As in previous activities, some pupils may identify that there are five without needing to count individually from 1. Ask: *Is the number of apples more than the number of children or less than the number of children?*
- Choose pupils to give out the cards and count aloud as they do so. Ask: *Do we have enough apples for the picnic? How many more do we need?*
- Continue the activity with other picnic items (concrete or pictorial), practising counting other numbers to 10.

• Work through Question 3 in the Practice Book, counting each group and writing the total. The image in the Practice Book links with the park theme.

(i) One-to-one correspondence is when one object (or image or symbol) is assigned one number and one number only. Young pupils often count objects more than once (or miss them out) and so it is important to give them opportunities to practise allocating one number to one item. For concrete counting, provide multi-sensory ways of ensuring each is counted once (for example dropping marbles into a bucket and counting a new number every time the sound is heard, or placing beads in an egg box one to a segment before counting them). For pictorial counting (as in Question 3 in the Practice Book), encourage pupils to point to each new picture as they count them or make a mark with a pencil to show that they have been counted.

Same-day intervention

• Provide pupils with groups of five objects in order to practise counting to five. Continue with plenty of concrete challenges (for example: *Can you attach four clothes pegs to this piece of cardboard?*) before moving on to pictorial counting to five.

Same-day enrichment

• Set pupils a reverse challenge where they are given the numbers to begin with and have to then draw the image. For example: *Can you draw a house with five windows, ten flowers outside, two doors, three clouds in the sky (and so on)?* Use pictures to set the challenge as per the table used to answer Question 3:

5	10	2	3

Challenge and extension question

Question 4

4 How will you colour the last three faces to continue the pattern?

This question tests pupils' abilities to identify repeating patterns, using counting to help. A useful way to help pupils to recognise the pattern and continue it is to provide several large versions of the three faces. Ask them to model the repeating pattern before deciding what the final three faces should be.

Chapter 1 Numbers up to 10

Unit 1.2
Let's sort (1)

Conceptual context

From infancy, children learn to sort and classify by organising their understanding of language, people and objects within their environment. Each new word and experience helps them to build up an understanding of how the different parts of their environment relate to themselves and to each other. This process of making sense of the environment is the first step in the important mathematical activities of matching, sorting and classifying. Pupils will have experience of these activities through play and sorting collections of everyday and small world resources. As they explore their environment, they notice how objects are alike and how they are different. They begin to sort them by criteria that have meaning to them: characteristics such as colour, size, shape and function.

This unit and the subsequent one help to formalise these skills. At this stage, pupils need to experience real objects, so during the lesson there should be an emphasis on the 'concrete' aspect of the Concrete Pictorial Abstract approach, with pupils being given opportunities to sort and classify a range of everyday objects.

The sorting process involves three interlinked steps:
- decide the characteristic to sort by
- physically sort the objects (or pictures of objects)
- explain the rationale for sorting.

The resources that pupils use for sorting do not need to be specifically mathematical. At this stage, it is the thought process involved that will enable pupils to develop their sorting skills. Pupils simply need repeated opportunities to practise sorting and justify their choices. These activities should encourage careful observation and discussion, and develop the use of precise language. Being able to articulate their thinking about sorting everyday objects will support conceptual development and give pupils linguistic competence and confidence that can later be readily transferred to wholly mathematical contexts.

Learning pupils will have achieved at the end of the unit

- Pupils will have explored the concept of sorting objects, and pictures of objects, into sets according to a rule or criterion (Q1, Q2)
- Pupils will have reinforced their ability to describe, in full sentences, characteristics, similarities and differences related to everyday objects (Q1, Q2)
- Pupils will have developed their ability to generalise about a set of objects and used conjecture and reasoning about the items to make decisions about any that do not fit (Q2, Q3)
- Pupils will have practised expressing their thinking in complete sentences to describe a property shared by all objects, with one exception (Q3)

Resources

small world toys (e.g. domestic animals, wild animals, sea creatures, vehicles); large sheets of paper (white and coloured); sets of natural objects (e.g. shells, leaves, fruit, vegetables); beads; buttons; coloured counters; coins; sets of objects made from different materials (e.g. plastic/metal/wood); 2-D/3-D shapes; interlocking cubes; hoops or sorting rings; catalogues to cut pictures from; **Resource 1.1.2 Can you see … ?**

Vocabulary

sort, sets, group, similar, same, different, criteria, characteristic, odd one out

Chapter 1 Numbers up to 10

Unit 1.2 Practice Book 1A, pages 3–4

Question 1

What learning will pupils have achieved at the conclusion of Question 1?

- Pupils will have explored the concept of sorting objects, and pictures of objects, into sets, according to criteria.
- Pupils will have reinforced their ability to describe, in full sentences, characteristics, similarities and differences related to everyday objects.

Activities for whole-class instruction

- Prior to the lesson, lay out an assortment of two different types of object, about ten in total, for example small world animals and vehicles. The objects should be easily divided into two sets.
- Ask pupils to look at the items and tell them that two different groups of objects have been muddled up. Challenge them to identify the two groups of objects. Can they give a name to each group? In the example suggested, names for the groups would be 'animals' and 'vehicles'.
- On a large piece of paper, draw two circles and write 'animals' and 'vehicles' on sticky notes to label them. Choose an individual pupil to pick an object to place in the correct circle and to explain their reasoning in a full sentence, for example: *The horse is a type of animal so it belongs in the animals group.*
- Pupils continue to place objects in the circles, ensuring that they use a full sentence as they place their chosen objects.
- Repeat this activity with different groups of objects, using a wide range of objects to ensure that pupils can confidently sort and group two sets of objects.
- Use the objects to model use of the words 'similar' and 'different', for example: *The sheep and donkey are similar because they are both animals.* Give pupils opportunities to practise using these words.

- Use the interactive whiteboard to show pictures of two groups of objects, for example about ten different flowers and birds. Explain that this time, instead of moving real objects, they should mark the pictures with different symbols to show which group each object belongs to, for example Group A (Flowers) and Group B (Birds).
- Again, as individual pupils identify an object to sort by marking it with the correct symbol, they should explain their reasoning in a full sentence, for example: *I have written A under this picture because it is a flower and flowers belong in Group A.*
- Work through Question 1 in the Practice Book, drawing a line to sort each object into the correct box.

Same-day intervention

- Provide pupils with two groups of three objects to sort so that they have a smaller number of objects on which to focus. Continue with plenty of concrete practice, gradually increasing the number of objects until pupils can cope with sorting 10 or 12 objects.
- Discuss with pupils how resources in the classroom are sorted by looking at trays of different toys, construction apparatus and so on. Talk about sorting when tidying up takes place. Can they spot items that have been put away incorrectly? This will continue to develop their sorting skills.

Same-day enrichment

- Ask pupils to suggest and justify additional items that would fit into groups using a full sentence, for example: *A lion is a type of animal so it could be added to the animal group.*
- Show pupils an object, for example a football, and ask a pupil to tell you the name of another object that is similar and would fit in the same group. Ask them to explain why. Ask more pupils to add to the group. They need to listen carefully to the first additional object because a single object may fit into different groups, for example football, tennis ball – types of ball; football, football boots – things required to play football.
- In pairs, pupils make their own assortment of two groups of concrete objects and another pair of pupils identifies and sorts the groups.
- Pupils prepare their own 'Sorting two groups' worksheet by cutting and sticking pictures from catalogues and magazines that can be sorted by others.

Chapter 1 Numbers up to 10

Unit 1.2 Practice Book 1A, pages 3–4

Question 2

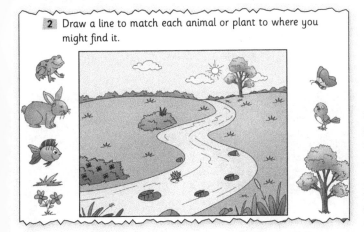

2 Draw a line to match each animal or plant to where you might find it.

What learning will pupils have achieved at the conclusion of Question 2?

- Pupils will have learned to sort objects into sets according to a rule or criteria.
- Pupils will be developing language to describe what is the same and what is different about objects.
- Pupils will have practised expressing their ideas in full sentences describing a general property that objects share.

Activities for whole-class instruction

- Remind pupils that you can sort or group objects according to a rule, for example objects that are red, creatures that can fly.
- Make a collection of models (or pictures) of vehicles that travel:
  - in the air, for example rocket, helicopter, aeroplane, hot-air balloon
  - on land, for example car, lorry, bicycle, bus
  - on water, for example canoe, speedboat, ferry, container ship.
- Ask pupils if they can suggest three groups that the vehicles could be sorted into. Encourage them to spot the criteria: land, air and water.
- Use three pieces of paper as shown to sort the models/pictures.

white paper for **air**	
green paper for **land**	blue paper for **water**

- These activities could be completed outdoors, using hoops or large sorting rings.
- Provide pupils with pictures from catalogues that show a selection of items from three different rooms in the house (for example kitchen, bathroom and living room). Ask them if they can sort them into three groups and explain their sorting.
- Work through Question 2 in the Practice Book, showing whether the plants and animals live in the air, on land or in the water.

Same-day intervention

- Work through **Resource 1.1.2** Can you see … ? In this activity, pupils must focus on particular attributes of objects and animals, and count objects with those attributes.
- Think about sorting collections of objects that are of particular interest to individual pupils to make the activity more personally relevant.

Same-day enrichment

- Challenge pupils to sort groups of mathematical items such as triangles, circles and rectangles or 2-D/3-D shapes.
- Provide pupils with pictures from catalogues and ask them to cut out and sort groups of objects. A camera may be useful to record their work.

Question 3

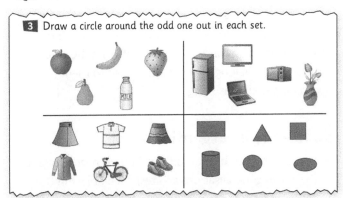

3 Draw a circle around the odd one out in each set.

What learning will pupils have achieved at the conclusion of Question 3?

- Pupils will have developed their ability to generalise about a set of objects and used conjecture and reasoning about the items to decide which one does not fit.
- Pupils will have practised expressing their thinking in complete sentences to describe a property shared by all objects, with one exception.

Chapter 1 Numbers up to 10   Unit 1.2 Practice Book 1A, pages 3–4

Activities for whole-class instruction

- Pupils have practised asking what is the same and what is different about groups of objects. The learning for this question requires them to find one object that does not fit with a group.
- Prepare sets of five or six concrete objects (or pictures) that are similar and add one that is different. Choose scenarios that are appropriate for your pupils. Possible ideas are:
  - 4–5 farm animals and 1 tractor
  - 4–5 vegetables and 1 fruit
  - 4–5 boats and 1 shark
  - 4–5 sets of shoes and 1 pair of socks
  - 4–5 fish and 1 crab
  - 4–5 identical birds flying and 1 perched.
- Encourage pupils to work in pairs to identify the group and recognise the one that does not belong through discussion.
- Choose a pupil to express this verbally, for example: *This is a group of (farm) animals. The one that does not belong is the tractor because it is not a (farm) animal.*
  Repeat what the pupil has said and introduce the term 'odd one out'. Say: *This is a group of (farm) animals. The odd one out is the tractor because it is not a (farm) animal.*
- Challenge pairs of pupils to use small world resources to make their own group of objects that includes an 'odd one out'. Invite pairs to identify and explain the odd one out.
- Allow pupils to work through Question 3 in the Practice Book, recognising the group and identifying the object that does not belong to the group. Listen to them explaining their reasoning.

Same-day intervention

- In order to find the 'odd one out', pupils need to make their own generalisations and test each object in the set provided against this. Therefore, initially make the object that does not fit very obvious, allowing pupils to focus on mastering the language.
- Model the sentence that describes the 'odd one out' and ask individual pupils to repeat it so that they gain confidence in articulating their reasoning. For the example of boats and one shark:

   This is a group of boats. The one that does not fit is the shark because it is not a boat. It is a fish.

Same-day enrichment

- Pairs make a desktop challenge using classroom objects including one object that does not belong to the group. Another pair find the odd one out and explain why.

Challenge and extension question

Question 4

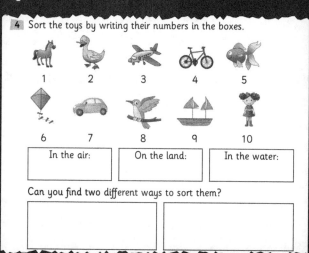

This question introduces pupils to the concept that objects can be sorted in different ways providing that the sorting criteria can be justified. The text hints that one way of sorting the objects would be into the same three categories as Question 2, namely those found in the air, on land and in water. The challenge is then to sort them another way into two groups, (possibilities are animals/not animals or types of transport/not types of transport). Encourage pupils to discuss their ideas with a talking partner.

A useful way to help pupils towards possible ways of sorting would be to have pictures of the toys (or the toys themselves) and hoops or sorting rings. Ask pupils to explain the criteria they have used to sort. This concept will be developed in the next unit.

Chapter 1 Numbers up to 10

Unit 1.3
Let's sort (2)

Conceptual context

Pupils already have considerable experience sorting collections of everyday and small world resources through play. They are building a bank of mathematical and everyday knowledge about how to sort different objects and how to use language in a precise manner to explain their reasoning. At this stage, they observe and describe a range of everyday characteristics as well as mathematical ones. Practice of this language will enable them to use these skills efficiently in the future to describe wholly mathematical sorting, for example odd/even numbers, multiples and so on.

In this unit, they learn that objects can be sorted in more than one way because each object possesses a number of different characteristics or properties. Aspects of appearance or function can be used to sort objects, for example size, colour, shape, gender. Pupils need to be able to identify possible groups and justify their sorting. A black kitten, for example could fit into a group of black objects, into a group of baby animals or into a group of animals with four legs.

Pupils need experience of grouping sets of objects (and pictures of objects) into groups in different ways. They can physically move similar objects so that they are close together or place them in sorting rings. They can draw a circle around each group or draw a line between the groups. They can mark items that belong in a group with the same symbol.

Learning pupils will have achieved at the end of the unit

- Pupils will have developed their visual perceptual skills (Q1)
- Pupils will have practised grouping sets of objects (and pictures of objects) according to a specific property or rule and explored showing the separate groups in different ways, for example physically placing them together, by circling each group, by drawing a line between the groups (Q1)
- Pupils will have learned that objects can be sorted in more than one way because an object has a number of different properties (Q1, Q2)
- Pupils will have reinforced their ability to explain in a full sentence the reason for the groups they have selected (Q1, Q2)
- Pupils will have consolidated their understanding and use of appropriate vocabulary to use in sorting activities, for example size, shape, property, similarity, difference (Q2)

Resources

sets of natural objects (e.g. shells, leaves, fruit, vegetables); beads; buttons; coins; everyday objects (e.g. kitchenware, stationery, clothing; 2-D/3-D shapes); interlocking cubes; coloured counters; dominoes; small world toys (e.g. domestic animals, wild animals, sea creatures, vehicles); sets of objects made from different materials (e.g. plastic/metal/wood); hoops; plates or sorting rings; **Resource 1.1.2** Can you see ... ?

Vocabulary

sort, set, group, object, similar, same, different, circle, size, criteria, characteristic, property

Chapter 1 Numbers up to 10

Unit 1.3 Practice Book 1A, pages 5–6

Question 1

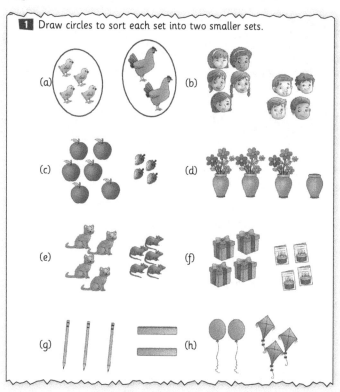

1 Draw circles to sort each set into two smaller sets.

(a) (b) (c) (d) (e) (f) (g) (h)

What learning will pupils have achieved at the conclusion of Question 1?

- Pupils will have practised grouping sets of objects (and pictures of objects) according to a specific property or rule and explored showing the separate groups in different ways, for example physically placing them together, by circling each group, by drawing a line between the groups.
- Pupils will have learned a number of properties that can be used for sorting.
- Pupils will have practised expressing in a full sentence the reasoning for the groups they have made.

Activities for whole-class instruction

- Remind pupils that you can sort objects according to similarities and differences.
- Prepare an assortment of objects (or pictures of objects) into groups, for example:
  - 3 red cubes and 4 green cubes
  - 4 vehicles with wheels and 3 vehicles with no wheels
  - 5 crayons and 2 whiteboard pens
  - 3 large triangles and 5 small ones
  - 4 exercise books and 3 mini whiteboards.

- Look at each assortment in turn and ask an individual pupil to explain in a full sentence how the mixture can be sorted, for example: *These can be sorted by colour into red cubes and green cubes.* Choose another pupil to carry out the sorting using hoops, plates or sorting rings. Finish by asking a third pupil to describe the two groups, for example: *This is a group of red cubes and this is a group of green cubes.* Repeat for the other assortments.
- Sometimes it is not possible to define a property such as colour or size. Pupils can easily sort exercise books and mini whiteboards because they differ in appearance and function, so in this case they might say: *They can be sorted into a group of exercise books and a group of mini whiteboards.*
- Show pupils a bowl of three apples and two bananas. Ask pupils to draw them in two separate groups on mini whiteboards. Share the diagrams they draw and discuss the different ways that they have shown them. Some pupils will draw them at either side of the whiteboard, others may separate the two groups with a line, while others may have drawn them in separate circles. Agree that all these representations are acceptable.
- Demonstrate the different ways by physically separating them, by drawing a line between them on a piece of paper and by placing them in sorting rings. Point out that sometimes they will be asked to show the groups in a particular way.
- Give pairs of pupils a selection of about ten buttons to sort and justify their sorting. The most likely method of sorting will be according to the number of holes as shown. Share and discuss what pupils do.

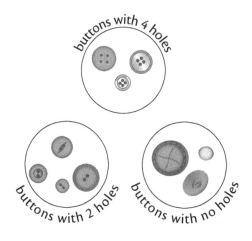

- Work through Question 1 in the Practice Book, sorting and circling the different groups. Ask individual pupils to describe their reasoning.

Chapter 1 Numbers up to 10

Unit 1.3 Practice Book 1A, pages 5–6

Same-day intervention

8			
	nine		
	7		

Give pupils a copy of this grid (from **Resource 1.1.2 Can you see … ?**) and ask questions to help them to focus on the features of things, for example:

- How many vehicles can you see? (1 – car)
- How many animals can you see? (5 – cow, kitten, dog and two sheep)
- How many pictures have two things in them? (2 – sheep and paint-brushes)
- How many pictures have black in them? (6 – dog, car, cow, kitten, sheep, open pencil case)
- Which two pictures are nearly the same? Explain what is different about them. (different possible answers)
- Describe the sheep (kitten and so on).
- What numbers are close to the pencil?
- Use 'spot the difference' pictures of appropriate challenge to help develop pupils' skills in observing and describing small differences. Follow this up by giving them identical pictures and challenging them to colour them so that there are five differences in the two pictures.
- Give pupils further practice sorting and describing assortments of two groups of objects. Model in full sentences the description of the two groups and ask pupils to repeat the sentence.

Same-day enrichment

- Observe and discuss how classroom resources are sorted for storage, for example small world toys, art materials or PE equipment. Encourage pupils to think about how they could be sorted in more than one way.

- Draw two large circles and label them. Challenge pupils to find objects that belong to the group. Possible pairs of labels are:
  - plastic objects/metal objects
  - farm animals/wild animals
  - clothes for hot weather/clothes for cold weather.

  Pupils should select their own two groups.

Question 2

What learning will pupils have achieved at the conclusion of Question 2?

- Pupils will have learned that the same object can be sorted in different ways because an object has a number of different properties.
- Pupils will have practised identifying suitable criteria to sort the same set of objects in different ways, for example size, colour, 2-D/3-D, and will have justified their choices.
- Pupils will have reinforced their ability to explain in full sentences the reason for the groups they have selected.
- Pupils will have consolidated their understanding and use of appropriate vocabulary to use in sorting activities, for example size, shape, property, similarity, difference.

Activities for whole-class instruction

- Prepare two identical muddled sets of ten interlocking cubes and counters, for example five interlocking cubes: three red, two blue and five counters: three red, two blue.

Chapter 1 Numbers up to 10 Unit 1.3 Practice Book 1A, pages 5–6

- Choose a pupil to sort one set physically into groups of cubes and counters. Ask another pupil to describe the groups, for example: *This is a group of cubes. This is a group of counters.*
- Ask them if they can see another way that the mixture can be sorted. Elicit from them that they can be sorted by colour into red things and blue things. Choose a pupil to sort the second set in this way and another pupil to explain, for example: *This is a group of red things. This is a group of blue things.*
- Ask: *Which way of sorting is correct?* Discuss the idea that both ways of sorting are correct; they are simply different. The way that things are sorted depends on what is required. If the requirement were for red things, then the first way would not be the way to sort the objects.
- Make a collection of large and small 2-D shapes. A possible selection is shown.

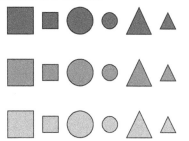

- Ask pupils to discuss with a talk partner some different ways that these objects could be sorted. Elicit that they can be sorted by colour, by size or by shape. Choose individual pupils to sort them in these different ways and other pupils to describe the groups, for example: *This group is large shapes and this group is small shapes.*
- Work through Question 2 in the Practice Book, sorting and circling the same set of objects so that they are grouped in different ways.

Same-day intervention

- Prepare an assortment of animal pictures, for example three large cats, three small kittens, three large dogs and three small puppies. Ask pupils to tell you what they can see in the pictures and to suggest ways that they can be sorted. Elicit that they can be sorted into cats and dogs or into young/small animals and adult/large animals.
- Ask a pupil to sort them into cats and dogs and another pupil to describe the two groups in a full sentence.
- Put the pictures together again and ask another pupil to sort them into young animals and adult animals. Ask another pupil to describe in a full sentence the two groups.
- Ensure that all pupils can confidently repeat the sentence describing the two groups.

Same-day enrichment

- Give pupils a full set of British coins and challenge them to find different ways of sorting them. Possible ways to sort them are small/large; circular/not circular; silver/not silver; copper/not copper. Don't expect pupils to know the value of coins yet and don't teach it now, but you will find that some pupils do already know and some sorting strategies might involve value. Pupils might sort coins into more than two groups or into overlapping/intersecting groups.

Challenge and extension question

Question 3

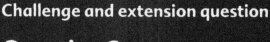

3 Draw a line to match each animal to the correct set.

Do any of the animals belong in both sets? Write the animals here.

In this question, pupils will learn that an object can belong to more than one group because it possesses a number of properties.

The diagram in the question shows overlapping or intersecting sorting circles. Objects that belong in both circles, that is, four-legged animals that climb trees, for example the squirrel, should be placed in the overlapping area, known as the intersection.

ⓘ This type of diagram is known as a Venn diagram, named after the English mathematician who first described them. At this stage, pupils do not need to know the vocabulary of Venn diagrams, only understand the logic of them.

Chapter 1 Numbers up to 10

Unit 1.4
Let's count (1)

Conceptual context

Pupils come to the classroom with vastly different mathematical experiences. These early lessons are important to lay sound mathematical foundations and to ensure that there are no gaps in individual mathematical knowledge. In this lesson, pupils learn to count in order to 10, accurately and reliably, not missing out any numbers or saying numbers in the wrong order. They learn to match this sequence of numbers with objects (one-to-one correspondence). Finally, they use this sequence of numbers to determine cardinal value, understanding that counting is the appropriate way to answer the question 'how many'.

At first, pupils count objects by physically moving each object they count or by touching each object. The next stage is to point to the objects closely, then to point less closely and finally to count 'by eye', simply fixing their gaze on each object in turn. It is easy to make errors when counting by eye and probably best at this stage to model counting by pointing.

The concept of conservation of number is the recognition that the quantity does not change when objects are physically rearranged. Some pupils may already be able to conserve number when they arrive in Year 1 – most others will achieve it during Year 1; they develop their ability to conserve number through experience of counting quantities that are arranged, counted, rearranged and counted again.

Pupils gradually develop number sense so that when they see or hear a number, for example the word 'five' or the symbol '5', they know that these both mean five countable things. They appreciate that 'fiveness' is independent of the way the objects are arranged or the order in which they are counted. At this point, the number takes on its own existence as a mental concept that can be thought about, thus opening the way for manipulating numbers.

Learning pupils will have achieved at the end of the unit

- Pupils will consistently be able to recognise numerals to 10, expressed as a numeral (2), as a word (two) or as a collection of objects (🚌 🚌) (Q1)
- Pupils will have consolidated counting accurately and reliably to 10 (Q1)
- Pupils will know that numbers increase sequentially one by one (Q1)
- Pupils will be developing their concept of conservation of number (Q1)
- Pupils will have begun to appreciate patterns of numbers, enabling them to subitise small numbers of objects, because it is easier, and to count larger numbers, because it is necessary (Q1)

Resources

mini whiteboards; counters or bricks; interlocking cubes; small world resources including fruit and vegetables; large numerals; set of 1–10 number cards (numerals and words, individual and class set); everyday objects; outdoor items (leaves, pebbles, twigs)

Vocabulary

count, match, numeral, number, one, two, three, four, five, six, seven, eight, nine, ten, 1, 2, 3, 4, 5, 6, 7, 8, 9, 10, how many … ? pattern

Chapter 1 Numbers up to 10

Unit 1.4 Practice Book 1A, pages 7–8

Question 1

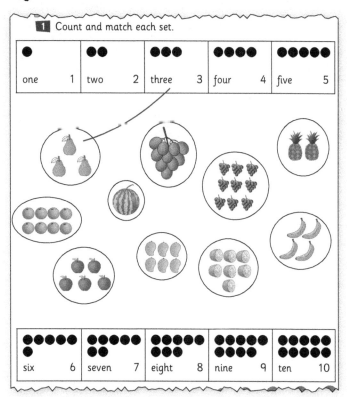

What learning will pupils have achieved at the conclusion of Question 1?

- Pupils will consistently be able to recognise numerals to 10, expressed as a numeral (2), as a word (two) or as a collection of objects (🚌 🚌).
- Pupils will have consolidated counting accurately and reliably to 10.
- Pupils will know that numbers increase sequentially one by one.
- Pupils will be developing their understanding of the concept of conservation of number.
- Pupils will have begun to appreciate patterns of numbers, enabling them to subitise small numbers of objects, because it is easier, and to count larger numbers, because it is necessary.

ⓘ Counting numbers, 1, 2, 3 … are known as **cardinal** numbers. A cardinal number answers the question 'how many?' When objects are counted, the total number of objects is the last number counted. Each object is counted only once using cardinal numbers. The order in which the objects are counted or how they are arranged does not matter.

👁 Look out for … pupils who are multilingual. The principles of one-to-one correspondence and cardinality are not dependent on language so encourage pupils to practise counting in their mother tongue. This will reinforce learning.

👁 Look out for … pupils who miss out numbers or consistently count in the wrong order. Take time to correct these errors.

Activities for whole-class instruction

- Give ten pupils number cards, 1–10, and ask them to put themselves in the correct order as quickly as possible. Repeat with cards that have the number words written on them. Count aloud with pupils forwards and backwards. Variations on this may include pegging numbers/number names on a washing line or arranging other sets of classroom numbers.
- Pick a number card at random and ask what number comes next, so is 1 more than, and what number comes before, so is 1 less than.
- Read counting books with pupils, for example *Handa's Hen* by Eileen Brown (Walker) or *Ten Terrible Dinosaurs* by Paul Stickland (Corgi).
- Choose five pupils to come to the front of the class. Ask how many children there are and establish there are five. Ask them to squeeze together very tightly and ask again how many there are. Ask them to spread apart and repeat the question.
- Repeat the exercise with other numbers using counters, bricks or small world toys to ensure that pupils understand that the number remains the same whether the objects are close together or spaced apart.

- In an outdoor setting, set pupils a series of challenges, for example: *Find 6 stones; 9 conkers; 7 leaves; 3 feathers.*

Chapter 1 Numbers up to 10

Unit 1.4 Practice Book 1A, pages 7–8

- Give pairs a plate and access to items for sorting, for example interlocking cubes, plastic minibeasts. Ask one pupil to place a specific number of those items, say eight, on their plate. They ask their partner to check the number. The choice of objects can be linked to the current curriculum topic. Change roles and repeat with other numbers. Each pair should keep score of how many times they get it right by putting a cube into a pot/bowl. How many cubes has the winning pair got in their pot/bowl after five minutes?

- Play a shopping game with small world fruit and vegetables. Choose one pupil to 'buy' the first item and put it in the 'trolley' (there is no need for money to be involved) say: *I went to the shop and I bought one apple.* Choose another pupil to make the next purchase, say: *I went to the shop and I bought one apple and two oranges.* Ask the class to close their eyes and repeat the sentence, so that they are counting incrementally and visualising the increasing numbers. Carry on adding more items to the trolley and repeating the shopping list, increasing by one each time until there are ten different items. Play variations on this game, for example meeting animals on a walk or relatives at a party.

- Give pupils mini whiteboards. Show them a number and ask them to draw that number of circles/crosses. Ask them to show their boards to check. Repeat with other numbers and shapes.

- Work through Question 1 in the Practice Book, counting and matching the pictures.

Same-day intervention

- It is important to build on the knowledge that pupils bring to the classroom to help them 'own' numbers. Show them the digit card '4' and ask what that number makes them think of. They may say that there are four people in their family, that their sibling is four years old or that a four-leafed clover is meant to be lucky. Repeat with other numbers. Ensure that each pupil contributes to the discussion. Generic ideas include:
  - 2 – 2 wheels on a bike; 2 arms, 2 legs, 2 eyes; twins
  - 3 – 3 sides on a triangle; 3 children in triplets; 3-legged stool; 3 poles in a tripod/wigwam
  - 4 – 4 legs on a chair; 4 legs on most animals; 4 corners on a square
  - 5 – 5 fingers on each hand; 5 arms on a starfish; 5 rings in the Olympic symbol
  - 6 – 6 legs on insects; 6 is the highest number on a dice; 6 eggs in a box (half-dozen)
  - 7 – 7 days in a week; 7 sides on 20p and 50p coins; 7 is many people's favourite number
  - 8 – 8 legs on a spider; 8 arms on an octopus
  - 9 – 9 a cat is said to have 9 lives; golf courses often have 9 holes
  - 10 – 10 fingers; 10 toes; 10 pins in a bowling alley

  Collect their ideas and add sticky notes and pictures to a working wall.

- Use the Internet to find interactive counting games for individual pupils for additional practice. There are many drag-and-drop type games readily available for practice.

- Sing and act number songs and rhymes, such as 'One elephant went out to play' or 'Ten in the bed'.

Same-day enrichment

- Let pupils choose a number as a focus, for example 6. Using paint and printing blocks, ask them to print sets of six (or another chosen number) printing blocks in different arrangements, trying to find as many ways as possible. Look at and discuss with partners the different patterns that they make. The possible patterns for each number will begin to be established in pupils' minds.

Repeat with other numbers. Pupils' prints could be displayed in the classroom alongside the corresponding numeral and number word.

- Give small groups of pupils a set of number cards, 1–10, and challenge them to set up their own outdoor counting display using only natural materials. They can use different leaves, daisies, dandelions, conkers, acorns, stones, twigs and so on. Encourage them to think about how they arrange the objects. Photograph their displays.

Chapter 1 Numbers up to 10   Unit 1.4 Practice Book 1A, pages 7–8

Challenge and extension question
Question 2

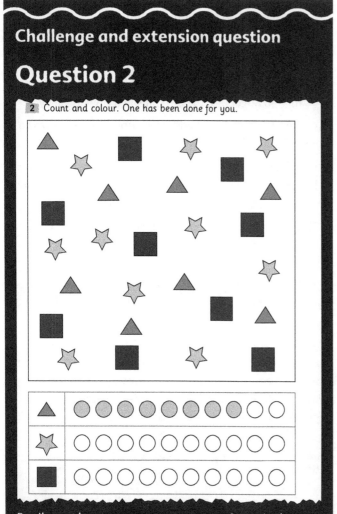

Pupils may be very competent at counting a single group of objects, but in this question three different sets of objects are arranged randomly, which makes the task considerably more challenging. Encourage pupils to scan across the objects from left to right, pointing and counting the chosen shape in a logical progression. Pupils need to retain the number they have reached until they locate the next shape that needs to be counted. Having found the total for each shape, they colour the correct number of circles on the ten-frame.

Chapter 1 Numbers up to 10

Unit 1.5
Let's count (2)

Conceptual context

Pupils' mathematical development requires the overlapping and gradual merging of verbal, cognitive and motor abilities. In this lesson, pupils continue to practise counting to answer the question 'How many?' with a new focus of learning to write the numbers 1–10, as words and numerals.

 Correct formation of numerals takes practice, extensive repetition and gradual development of fine motor control.

A good introduction to formal recording is for pupils to have experience of tracing numbers by asking them to use the index finger of their dominant hand to trace the number carefully while you give spoken instructions. (You could explain 'tracing' as drawing an invisible numeral.) It is important to stress the starting point and direction of lines as well as when to keep their finger on the page and when to lift it off. Demonstrate by tracing large numerals on the board so that pupils can see what you are describing. Repeat each numeral many times before moving on to the next.

1: To trace the number 1, start at the top and draw straight down.

2: To trace the number 2, start at the top and follow the curve. Without lifting the pencil, draw straight along the bottom.

3: To trace the number 3, start at the top and follow the two curved lines without lifting the pencil.

4: Tracing the number 4 requires two strokes. Start at the top, draw diagonally to the left and then horizontally to the right. Start the second stroke at the top and draw down, crossing the middle of the horizontal line. Start the second stroke at the top-left corner and draw horizontally to the right.

5: Tracing the number 5 requires two strokes. Start at the top-left corner, draw down about half the height of the finished number. Without lifting the pencil, draw around the curve.

6: To trace the number 6, start at the top, draw down the curve and finish with a loop.

7: To trace the number 7, start at the top and draw along the top, then draw a diagonal line down to the bottom.

8: To trace the number 8, start at the top and draw a letter S. Without lifting the pencil, carry on drawing a backward S to meet back at the top. Discourage pupils who draw two circles.

9: To trace the number 9, start at the top right and draw around the curve, straighten the line as you go back up and, without lifting the pencil, draw a straight line down to the bottom.

Once they have confidence in tracing numbers, pupils can begin to write them. It is important to provide opportunities to write numbers in different and fun ways. These informal contexts will make the learning process more enjoyable and relaxed so that pupils do not become anxious about making mistakes. Writing numerals does not need to be done in order. Pupils generally find 1 and 7 quite easy, while 8 and 5 are the most challenging.

 Pupils will also be learning to read and spell the number words. Ten, as a CVC word, is very simple to spell. The spelling of the number words, five, six, seven, nine and three, is covered in the Year 1 National Curriculum spelling guidance. The remaining numbers may be part of your spelling programme.

As pupils engage in counting activities, they begin to recognise common patterns of numbers, for example the arrangement of the spots on dominoes and dice or increasing numbers within ten-frames. These visual patterns are important aspects of their number concepts, becoming tools for thinking. At this early stage in pupils' mathematical development, the visual images of these patterns will support their counting, making it quicker.

Learning pupils will have achieved at the end of the unit

- Pupils will be able to write the numerals from 1 to 10 correctly (Q1)
- Pupils will have practised spelling the number words for 1–10 (Q1)
- Pupils will have had further practice counting accurately up to ten objects (Q1)
- Pupils will have continued to develop their concept of conservation of number (Q1)
- Pupils will have developed their knowledge of patterns of numbers, enabling them to subitise small numbers of objects, because it is easier, and to count larger numbers, because it is necessary (Q1)

Resources

mini whiteboards; counters; interlocking cubes; large numerals; sets of 1–9 digit cards (individual and class set); everyday objects; outdoor items (leaves, pebbles, nuts); number guides; large foam or plastic numbers; sandpaper numbers; sand tray; dominoes; ten-frames

Vocabulary

count, match, numeral, number, symbol, one, two, three, four, five, six, seven, eight, nine, ten, 1, 2, 3, 4, 5, 6, 7, 8, 9, 10

Chapter 1 Numbers up to 10

Unit 1.5 Practice Book 1A, pages 9–11

Question 1

1 Count each set and then draw a line to match it to the correct number and number name.

1	2	3	4	5
one	two	three	four	five

Colour the correct number of dots to show the same number of balls. Write the correct number name and numeral next to each set.

balls	dots	name	numeral
●●	●●○○○	two	2
●●●	○○○○○		
🏈	○○○○○		
●●●●●	○○○○○		
●●●●	○○○○○		

Count each set and then draw a line to match it to the correct number and number name.

6	7	8	9	10
six	seven	eight	nine	ten

Colour the correct number of dots to show the same number of flowers. Write the correct number name and numeral next to each set.

flowers	dots	name	numeral
🌼🌼🌼🌼🌼🌼🌼🌼	●●●●●●●●○○	eight	8
🌸🌸🌸🌸🌸🌸🌸	○○○○○○○○○○		
🌷🌷🌷🌷🌷	○○○○○○○○○○		
🌺🌺🌺🌺	○○○○○○○○○○		
🌼🌼🌼	○○○○○○○○○○		

What learning will pupils have achieved at the conclusion of Question 1?

- Pupils will be able to write the numerals from 1 to 10 correctly.
- Pupils will have practised spelling the number words for 1–10.
- Pupils will have had further practice counting accurately up to ten objects.
- Pupils will have continued to develop their concept of conservation of number.
- Pupils will have developed their knowledge of patterns of numbers, enabling them to subitise small numbers of objects, because it is easier, and to count larger numbers, because it is necessary.

Activities for whole-class instruction

- Pupils practise tracing numbers using the forefinger of their dominant hand:
  - over large foam or plastic numbers
  - on worksheets.
- Practise drawing numbers correctly.

1 2 3 4 5
6 7 8 9 10

  - with a paintbrush and water in the playground
  - with a finger on a friend's back
  - with a finger in a tray of sand
  - with a pen on a mini whiteboard
  - with coloured chalk in the playground
  - with glue and glitter
  - with shaving cream.

- Give pupils mini whiteboards and set out a number of counters or objects. Choose a pupil to count them. Ask pupils to write the number on their whiteboard. Repeat with other numbers.

- Help pupils to prepare a chart as shown and to complete it with buttons, counters or 1p coins. Arranging the objects in this way will help pupils to see how counting numbers increase one at a time.

21

Chapter 1 Numbers up to 10 Unit 1.5   Practice Book 1A, pages 9–11

- Work through Question 1 in the Practice Book, matching the objects with the correct number and numeral. Encourage pupils to complete all circle patterns by filling the top row of 5 from the left, before starting to fill the bottom of 5 from the left as shown in the example.

  Look out for … left-handed pupils may need additional support in forming numerals, as they cannot easily see what they are writing. They may find it easier to write an 8 in reverse. Similarly, pupils with dyspraxia will find this challenging. Ensure that such pupils are allowed to demonstrate their true cognitive abilities and are not penalised for immature motor control. These pupils should be receiving support to help with handwriting and as their handwriting skills improve, so will their ability to write clear numerals.

- Reversals of numbers are common at this stage of development. Most pupils will improve with time and practice. Give pupils visual and verbal cues as shown and described to aid the correct formation.

Same-day intervention

- Provide worksheets to practise drawing numerals that use arrows to show the correct direction required for each stroke. Check which numerals individual pupils find difficult and provide focused support.
- Write numbers on white paper using a pink or blue highlighter pen. Give pupils a yellow highlighter and ask them to trace the numbers carefully. If they trace accurately, the colour will change from pink to orange, or from blue to green.
- Use PE small apparatus, for example beanbags, quoits, or spot markers, for pupils to play counting games. Place the apparatus a sensible distance away. Give pairs of pupils a hoop in which to place items and then challenge them to run and collect a number of objects, for example six beanbags. Choose a pupil to set a new challenge.

Same-day enrichment

- Give pupil pairs 30 blank playing cards and ask them to make three sets of cards for the numbers 1–9, showing numerals, pattern of dots and number words, for example 4, ∷ and **four**. These can be used for sorting or snap games and are a useful resource for future lessons.

- Show pupils a number, for example 7. Ask them what number is 1 more than 7. Elicit the answer 8. Model this with counters if necessary. Repeat with other numbers.
- Show pupils a number, for example 6. This time, ask them what number is 1 less than 6. Elicit the answer 5. Model this with counters if necessary. Repeat with other numbers.
- Prepare photos of popular car logos.

  Let pupils look at and talk about the logos in pairs or small groups. Ask them what ways they can think of to sort the logos. Possible suggestions are:
  - name of brand/no brand name
  - colour detail/only silver chrome
  - circle/no circle.

  Pupils whose parents have a different make of car may be able to bring in a photo of the logo the next day and see how it would fit into the different sorting criteria.

Challenge and extension question

Question 2

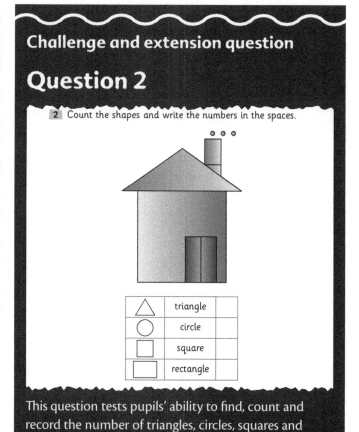

This question tests pupils' ability to find, count and record the number of triangles, circles, squares and rectangles within a diagram. Point out to pupils that the same shape may vary greatly in size.

Chapter 1 Numbers up to 10

Unit 1.6
Let's count (3)

Conceptual context

In this unit, pupils are consolidating their abilities to sort and count objects in more than one way by sorting and counting 3-D and 2-D shapes. The 3-D shapes illustrated in the Practice Book questions are cubes, cuboids, cylinders, triangular prisms and spheres. The 2-D shapes shown are circles, squares, rectangles, triangles and trapeziums.

At this stage, pupils are simply expected to be able to identify similarities and differences in shapes, they are not expected to know the names of the shapes. The names and associated vocabulary will be learned when conceptual knowledge about shapes is introduced in Chapter 4. Should any challenging questions from pupils arise during this unit, refer to Chapter 4 for guidance.

While working on the unit, model the correct vocabulary, ensuring that you are showing pupils an example of the shape you want them to sort. For example, to make a set of cylinders, say: *Look at the whole group of shapes, which ones are shaped like this one (hold up a cylinder)? How do you know they are similar, that they are cylinders?* Pupils may tell you that they can roll or slide (or cannot) or show you that they all have circles (circular faces). Each object has a number of different properties including shape (and, for 3-D objects, shape of faces), size and colour. These properties can be used to sort a collection of objects in a variety of ways. Give pupils time to explore 3-D and 2-D shapes informally, using them to make patterns and build structures.

Learning pupils will have achieved at the end of the unit

(In general, this unit expands the sorting and counting skills that pupils have developed in previous units by applying them in the context of shapes.)

- Pupils will have practised grouping 3-D shapes (and diagrams of 3-D shapes) according to a specific property or rule and explored showing the separate groups in different ways, for example physically placing them together, circling each group, drawing a line between the groups (Q1)
- Pupils will have learned that 3-D and 2-D shapes can be sorted in more than one way because shapes have a number of different properties (Q2)
- Pupils will have developed their ability to perceive similarities and differences in the context of 3-D and 2-D shapes (Q1, Q2)
- Pupils will have developed their visual perception skills of 3-D shapes (Q1, Q2) and 2-D shapes (Q2)
- Pupils will have reinforced their ability to explain in a full sentence the reason for the groups they have selected (Q1, Q2)
- Pupils will have been introduced to the names for 3-D shapes (cube, cuboid, sphere, cylinder, triangular prism) and names for 2-D shapes (square, rectangle, circle, triangle, trapezium) but are not yet expected to know the words themselves (Q1, Q2)

Resources

mini whiteboards; 3-D shapes (cubes, cuboids, cylinders, triangular prisms and spheres); 2-D shapes (circles, squares, rectangles, triangles and trapeziums); interlocking cubes; sorting rings; plates; 1–9 digit cards

Vocabulary

sort, count, similar, same, difference

Chapter 1 Numbers up to 10

Unit 1.6 Practice Book 1A, pages 12–13

Question 1

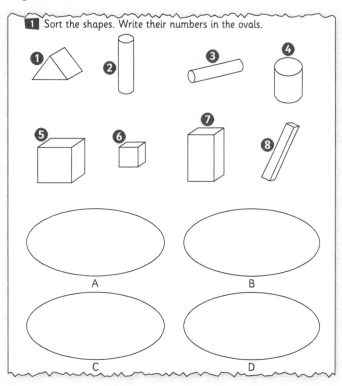

What learning will pupils have achieved at the conclusion of Question 1?

- Pupils will have practised grouping 3-D shapes (and diagrams of 3-D shapes) according to a specific property or rule and explored showing the separate groups in different ways, for example physically placing them together, circling each group, drawing a line between the groups.
- Pupils will have developed their ability to perceive similarities and differences in the context of 3-D shapes.
- Pupils will have reinforced their ability to explain in a full sentence the reason for the groups they have selected.
- Pupils will have been introduced to the names for 3-D shapes (cube, cuboid, sphere, cylinder, triangular prism) and associated vocabulary (face, square, rectangle, circle, triangle) but are not yet expected to know the words themselves.

Activities for whole-class instruction

- Show pupils a collection of red cubes of two different sizes, for example three small cubes and two larger cubes.

Ask how many different shapes there are in the collection. Establish that there is only one shape; they are all cubes, but they are not the same size.

Now show them a collection of red cylinders, for example two narrow/thin cylinders, two wide cylinders and one short cylinder.

Ask how many different shapes there are in this collection. Establish that again they are the same shape, called cylinders, but that they differ in size. Ask pupils to describe the similarities and differences. They may make observations such as: *They are all the same colour and they all have circles at the ends. They are not exactly the same because these two are longer than this one and these two are thinner than the others.*

Put both collections into a tray and mix them up. Choose individual pupils to pick one object at a time and place them in two sorting hoops, cubes and cylinders. Ask them to explain why they are placing their object in that sorting hoop.

- Start with a collection of cuboids and triangular prisms that have the numbers written on them.

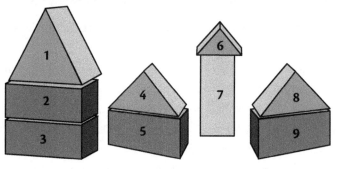

Give pupils mini whiteboards. Ask them to draw two sorting hoops and write the numbers of the cuboids in one hoop and the triangular prisms in the other. Check their recording.

- Complete Question 1 in the Practice Book, sorting the different groups. Ask individual pupils to describe their reasoning.

Same-day intervention

- Show pupils a set of three shapes where two shapes are the same and one is odd, for example two spheres and one cylinder or two squares and one triangle. Ask individual pupils to explain in a whole sentence which is the odd one out. Repeat with different sets of three objects to build confidence.

Chapter 1 Numbers up to 10

Unit 1.6 Practice Book 1A, pages 12–13

Same-day enrichment

- Give pairs or small groups of pupils two sorting hoops. Prepare a selection of 2-D shapes that includes circles and semicircles, plus some squares, rectangles, triangles and trapeziums. Review their understanding of the terms 'straight' and 'curved' by asking pupils to draw a straight line in the air, then a curved line. Challenge them to sort the shapes into those with straight lines and those with curved lines. As the semicircles belong in both sets, pupils need to deduce that an intersection is required.

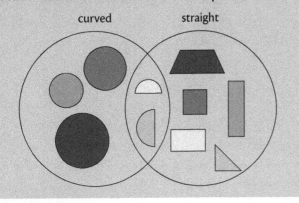

What learning will pupils have achieved at the conclusion of Question 2?

- Pupils will have developed their ability to perceive similarities and differences in the context of 3-D and 2-D shapes.
- Pupils will have practised grouping 2-D and 3-D shapes (and diagrams of 2-D and 3-D shapes) according to a specific property or rule and explored showing the separate groups in different ways, for example physically placing them together, circling each group, drawing a line between the groups.
- Pupils will have learned that 3-D and 2-D shapes can be sorted in more than one way because shapes have a number of different properties.
- Pupils will have reinforced their ability to explain in a full sentence the reason for the groups they have selected.
- Pupils will have been introduced to the names for 3-D shapes (cube, cuboid, sphere, cylinder, triangular prism) and names for 2-D shapes (square, rectangle, circle, triangle, trapezium) but are not yet expected to know the words themselves.

Question 2

Activities for whole-class instruction

- Prepare a selection of cylinders and cuboids, as shown.

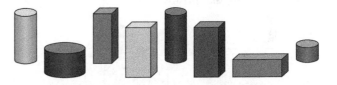

Ask how many different shapes there are in this collection. Establish that there are only two different shapes, cuboids and cylinders. Hold up an example of each shape as you name it. Choose some pupils to sort the shapes. Check that pupils understand and can explain the sorting.

Mix the shapes up again and ask if they can see another way that they could be sorted. Elicit that they could also be sorted by colour. Resort the shapes by colour.

- Give pupils access to a large selection of 2-D shapes: triangles, circles, squares and rectangles. Ask pairs of pupils to choose 8–10 shapes and arrange them to make an imaginary vehicle such as the one shown below.

Chapter 1 Numbers up to 10

Unit 1.6 Practice Book 1A, pages 12–13

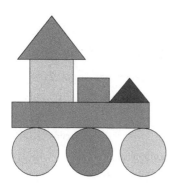

Ask another pair to sort the shapes in the vehicle by shape and then by colour.

- Give groups of pupils a collection of the same colour 2-D and 3-D shapes, for example one sphere, two circles, two cylinders, two cubes, one cuboid, one square, one rectangle. Ask them to sort the shapes in different ways. Encourage pupils to sort them into 2-D and 3-D shapes, into shapes with straight lines only and shapes with curves.

- Before pupils attempt Question 2, you should agree the colours that they will count and shade the left-hand boxes in the table accordingly. Then, work through Question 2 in the Practice Book, first counting the objects by colour/shade and completing the table, and then counting the objects by shape and completing the table.

Same-day intervention

- Prepare a simple selection of six shapes: three cubes and three triangular prisms.

Ask pairs to look at the shapes and explain how they could sort them. Ask them to complete the spoken sentences:
I can sort the objects by colour. There are … (2 red shapes, 2 blue shapes, 1 green shape and 1 yellow shape).
I can sort the objects by shape. There are … (3 cubes and there are 3 triangular prisms/shapes with triangular faces/shapes that have triangles).

Same-day enrichment

- Begin by asking pupils to carry out the intervention task above. Show them that the sentence: *I can sort the objects by colour: there are 2 red shapes, 2 blue shapes, 1 green shape and 1 yellow shape* can be expressed as: Red – 2, blue – 2, green – 1, yellow – 1. Work out the expression when the objects are sorted by shape.

- Now give pupils a larger collection of three different 3-D shapes that can be sorted in more than one way.

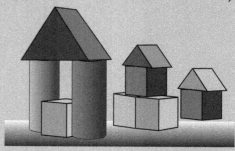

Explain to pupils that the shapes were sorted in two different ways but the labels have been lost. Can they work out the missing labels?

First way of sorting:

_____ – 4, _____ – 3, _____ – 2,

_____ – 1.

Second way of sorting:

_____ – 5, _____ – 3, _____ – 2.

If they need support, ask them to count how many different colours are in the collection (4). This should help them to appreciate that the first way of sorting is by colour.

Challenge and extension question

Question 3

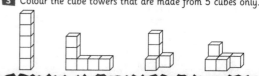

3 Colour the cube towers that are made from 5 cubes only.

This question tests pupils' abilities to interpret 2-D representations of 3-D arrangements of cubes. In 1 and 2, every cube is visible but diagrams 3 and 4 each have one cube that is wholly invisible. To help pupils learn to interpret 3-D diagrams, make sure that they have plenty of opportunities to handle bricks, cubes and other types of construction apparatus.

Chapter 1 Numbers up to 10

Unit 1.7

Let's count (4)

Conceptual context

In this unit, pupils are exploring the importance of seeing patterns of numbers to 10 as 5 plus 1, 5 plus 2 and so on, and that 10 is two groups of 5. Five-wise patterns on a ten-frame are practised where the two rows are successively filled. This leads to understanding the ease and usefulness of counting in fives and can include finger counting. Pupils continue to develop their ability to assign a number to standard spatial configurations, such as those on dice and dominoes – or fingers.

Although there will be opportunities to use manipulatives, increasingly, pupils will practise counting ten objects from a picture of a much larger selection of objects, randomly arranged, rather than counting physical objects. This means that they are no longer able to count objects by physically moving them. They need to develop strategies to count systematically and accurately that do not rely on concrete manipulation. Discuss and share the ways they use, for example counting from left to right or top to bottom. Most pupils of this age will still point as they count, which means that counting left to right may be slightly harder for left-handed pupils.

Learning pupils will have achieved at the end of the unit

- Pupils will have further practised accurately counting to 10 (Q1)
- Pupils will have continued to practise writing numerals correctly (Q1)
- Pupils will have reinforced their understanding that counting numbers increase sequentially one by one (Q2)
- Pupils will have begun to develop an understanding of 'same value, different appearance' (Q2)
- Pupils will have continued to develop visual images of the numbers 5–10, represented as 5 plus 0, 5 plus 1, 5 plus 2 and so on (Q2)
- Pupils will have had opportunities to confirm that two groups of five objects gives a total of ten objects, and committed this number fact to memory (Q3, Q4)
- Pupils will know there are five fingers on each hand and ten fingers on two hands (Q3)
- Pupils will have practised counting ten objects from a larger number of objects that are arranged randomly (Q4)

Resources

individual and class sets of 1–10 number cards; ten-frames; mini whiteboards; counters; interlocking cubes; sorting rings; plates; dominoes; bead strings; dice; 1p coins; small world toys; **Resource 1.1.7a** On the farm; **Resource 1.1.7b** Is this 10?

Vocabulary

count, whole, part, partition, ten-frame

Chapter 1 Numbers up to 10

Unit 1.7 Practice Book 1A, pages 14–15

Question 1

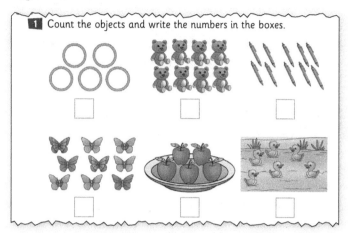

- Complete Question 1 in the Practice Book, counting the objects and writing the numbers in the boxes. Remind pupils to think about forming numerals correctly.

Same-day intervention

- Give pairs of pupils a selection of dominoes and ask them to find all the dominoes with six spots. The choice and number of dominoes in the selection can vary with the level of challenge.
  Repeat this for other target numbers.

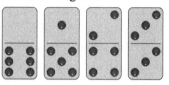

What learning will pupils have achieved at the conclusion of Question 1?

- Pupils will have further practise accurately counting to 10.
- Pupils will have continued to practise writing numerals accurately and correctly.

Activities for whole-class instruction

- Give pupils a plate and pile of counters, or other objects. Show a number, for example 7, and ask them to count that number on to the plate. Ask them to check one another's plates. Repeat with other numbers, sometimes simply saying the number and sometimes showing a number card.
- Provide pupils with mini whiteboards and say or show a number. Ask them to draw that number of circles or triangles.
- Show pupils a number of objects (or a picture of objects), sometimes arranged randomly and sometimes in an organised pattern. Ask them to write the number on their mini whiteboards. Repeat with other numbers.

Same-day enrichment

- Ask pupils to discuss whether it is easier to count numbers that are arranged randomly or in an organised pattern. Look at the two arrangements of 8 above. Establish that organised patterns are easier. Ask them to draw the number patterns that they recognise without counting. They may draw dice patterns or doubles. Share patterns with the group. These could be displayed on a maths working wall.

Question 2

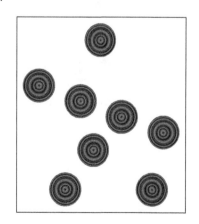

What learning will pupils have achieved at the conclusion of Question 2?

- Pupils will have reinforced their understanding that counting numbers increase sequentially one by one.
- Pupils will have begun to develop an understanding of 'same value, different appearance'.
- Pupils will have continued to develop visual pictures of the numbers 5–10, represented as 5 plus 0, 5 plus 1, 5 plus 2 and so on.

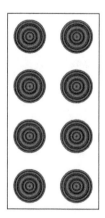

Chapter 1 Numbers up to 10

Unit 1.7 Practice Book 1A, pages 14–15

Activities for whole-class instruction

- Help pupils to use their fingers to count together from 5 to 10. Encourage pupils to think that 6 is 5 and 1 more, and so on.
- Ask pupils to prepare strings of ten beads: five white and five red (or give them prepared bead strings). Ask them to count seven beads on the string and show you.

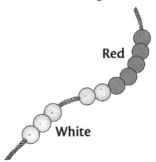

Repeat with other numbers.
Ask them to show you five beads and to start at 5 and count on.
- Complete Question 2 in the Practice Book. Try to listen to pupils as they read the numbers and ask them to explain how the pattern grows.

Same-day intervention

- Give pupils a ten-frame and cubes. Ask pupils to place five cubes in the top row and count them together. Ask them to show other numbers from 6 to 10.

Same-day enrichment

- Show pupils two ten-frames. Ask pupils which is the bigger of the two numbers. (B is a bigger number than A.) Can they explain that they know B is bigger because there are more cubes (more ones)? Play the following game in pairs.

A					
	*	*	*	*	*
	*	*			

B					
	*	*	*	*	*
	*	*	*	*	

In pairs, pupils place cubes in the ten-frame showing a number from 4 to 10.
Toss a coin – heads the pupil with the bigger number wins a point; tails, the pupil with the smaller number wins a point. The first person to score 5 points is the winner.

Question 3

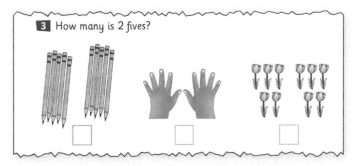

3 How many is 2 fives?

What learning will pupils have achieved at the conclusion of Question 3

- Pupils will have had opportunities to confirm that two groups of five objects gives a total of ten objects, and committed this number fact to memory.
- Pupils will know there are five fingers on each hand and ten fingers on two hands, and that two sets of 5 is 10.

Activities for whole-class instruction

- Ask pupils to count the fingers on one hand and then on the other. Establish that there are five fingers on each hand. Now count the total number of fingers on both hands to grasp that two groups of 5 makes 10.
- Ask pupils in pairs to each take five cubes and put the two sets of cubes together in a pile. Ask them to count the cubes and establish that every pair has a total of ten cubes. Some pairs could use small world objects or counters instead of cubes.
- Give pupils mini whiteboards and instruct them to draw a row of five circles. Then ask them to draw a second row of five circles underneath. Ask them to count the total number of circles and confirm that there are 10.
- Ask pupil pairs to set out two groups of five 1p coins in an organised way. Share the different patterns and check that each set does make 10. Possible groupings that pupils may choose include the following arrangements.

Chapter 1 Numbers up to 10

Unit 1.7   Practice Book 1A, pages 14–15

- Complete Question 3 in the Practice Book. Ask pupils to explain: *Two groups of 5 is 10.*

Same-day intervention

Work through **Resource 1.1.7a**, On the farm.

Pupils will count animals in the picture of a farm scene. They will finish by drawing a given number of birds onto the scene.

Same-day enrichment

- Ask pupils to compose collections of two groups of 5 or to draw them. Photograph and display them.

Question 4

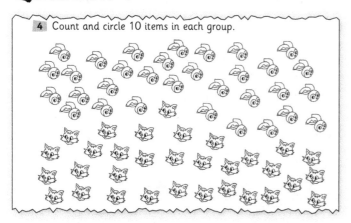

What learning will pupils have achieved at the conclusion of Question 4?

- Pupils will have practised counting ten objects from a larger number of objects that are arranged randomly.

Activities for whole-class instruction

- Ask pupils to count 10 from a much larger group of outdoor objects, for example acorns – allow them to point and move the objects. Repeat with other objects such as stones, leaves or conkers.
- Show pupils **Resource 1.1.7b** Is this 10?

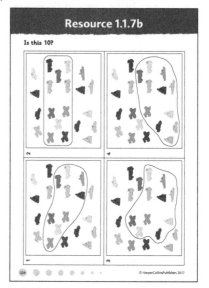

These are pictures in which pupils have tried to circle ten objects from a much larger group of objects. Show them each picture quickly and ask them to say if they can tell without counting that one of pupils definitely circled 10. Do they think the others are 10 or not?

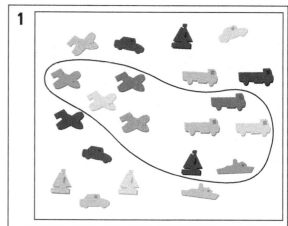

Chapter 1 Numbers up to 10

Unit 1.7 Practice Book 1A, pages 14–15

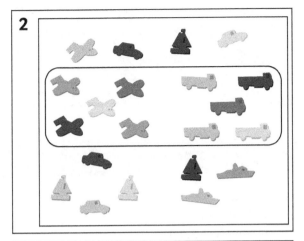

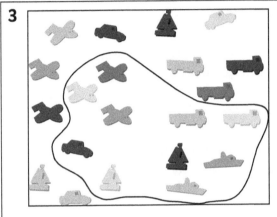

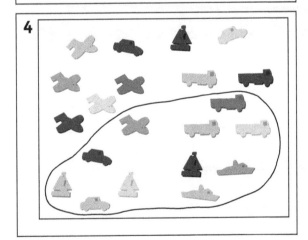

Pupils may spot that Picture 2 has 10 circled in two groups of 5 that are arranged in the familiar dice pattern for five. Establish that 10 is too big a number to subitise and that careful counting is required, unless objects are in a familiar pattern. In fact, Picture 1 has nine objects circled, Picture 3 has ten objects and Picture 4 has 11 objects circled. Check by counting.

- Complete Question 4 in the Practice Book. Ask pupils whether one group was easier to count. Listen to the answers, which may resemble: *I could circle the two hands without even counting because I **know** that each hand has five fingers and I **know** two lots of 5 make 10. I had to count the others very carefully because they were arranged in a random way.*

Same-day intervention

- Take pupils to the school dining room and look at the cutlery and crockery. Ask them to count ten knives, ten forks, ten spoons, ten plates, ten bowls, ten mugs. They may be able to use them to lay a table.

Same-day enrichment

- Gertie the Gardener has a new commission! Her job is to design a '10' garden. In the garden there are five flowerbeds and each bed must have ten plants in it. Ask pupils, individually, in pairs or small groups as appropriate, to draw a plan for the garden. Suggest to pupils that plants can be represented by small crosses or circles so that they do not become too bogged down in drawing and colouring.

  Share the designs and ask pupils to spot beds in each other's plans with similar patterns. Ask pupils to describe the layout of particular flower beds in whole sentences. If time permits, pupils could 'make' the garden from the plan using cubes for plants.

Challenge and extension question

Question 5

5 Count the cubes. Write the number in the box.

☐ cubes

This question continues to build pupils' abilities to interpret 2-D representations of 3-D arrangements of cubes. In the diagram, several cubes are wholly invisible, one cube in the middle layer and three in the bottom layer. To help pupils' ability to interpret 3-D diagrams, give them plenty of opportunities to handle bricks, cubes and other types of construction apparatus. There are 10 cubes in the diagram.

Chapter 1 Numbers up to 10

Unit 1.8
Let's count (5)

Conceptual context

The new learning in this unit is the introduction of 0 (zero). Zero is a number that means no objects are present. The number zero is one less than 1 and marks the start of a (positive) number line. Zero is both a number and the numerical digit (0) used to represent that number. As a digit, 0 is used as a placeholder to show that there are 0 units of that (column) value in that position.

While humans have always understood the idea of having nothing or no amount, they have not always used a symbol to represent zero. The Roman counting system had no zero. The symbol '0' originated in India around 800BCE but did not reach the West for several hundred years. In the 12th century, the mathematician Fibonacci wrote that zero can be used as a 'placeholder' to separate columns of figures.

Look out for … pupils who use the letter 'O' to describe zero. This is commonly used in everyday speech, for example O O 7 for James Bond. Point out that 'O' is actually a letter and always model the correct use of zero.

Learning pupils will have achieved at the end of the unit

- Pupils will have learned that 0 is a number that comes before 1 (Q1)
- Pupils will have learned the vocabulary associated with 0: zero, nothing, nil, nought, none (Q1, Q2, Q3)
- Pupils will have continued to practise counting up to ten objects in order to answer the question 'How many?' (Q1)
- Pupils will have further practised counting to 10 and writing numerals accurately and correctly (Q2)
- Pupils will have continued to develop an understanding of 'same value, different appearance' (Q2)
- Pupils will have continued to develop their ability to subitise small values and to recognise numbers that are arranged in standard spatial configurations (Q2)
- Pupils will have further practised drawing the correct number of objects for a given number (Q3)

Resources

interlocking cubes; red/blue counters; 1–10 dice; blank 10-sided dice; blank hexagonal spinner; number lines; blank number lines; fruit bowl; pasta shapes; online timer; drum; fruit or pictures of fruit; 0–10 number cards

Vocabulary

zero, nothing, nil, nought, none

Chapter 1 Numbers up to 10

Unit 1.8 Practice Book 1A, pages 16–17

Question 1

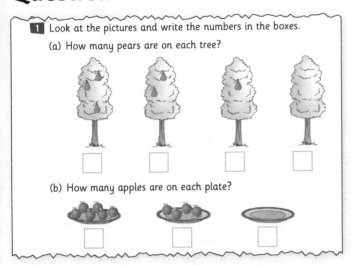

What learning will pupils have achieved at the conclusion of Question 1?

- Pupils will have learned that 0 is a number that comes before 1.
- Pupils will have learned the vocabulary associated with 0: zero, nothing, nil, nought, none.
- Pupils will have continued to practise counting up to ten objects in order to answer the question 'How many?'

Activities for whole-class instruction

- Practise counting backwards from 10 to 0 in a variety of different ways, for example counting down a line of ten pupils, as you touch one ask them to sit down.

 Ten children standing, nine children standing, eight children standing … zero children standing.

- Give pupils interlocking cubes to make number towers, from 1 to 10, in a staircase pattern. Each tower should be one more than the previous one.

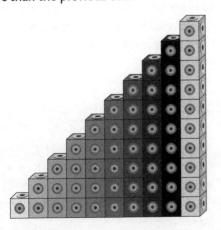

Use a random number generator (0–10) to give numbers. Ask pupils to point to the correct step on the staircase and then point and count down the staircase to 0.

- Show pupils incomplete number tracks and ask them to tell you the missing numbers.

0		2		4		6	7		9	10
	1		3	4	5			8	9	

- Show pupils the number 1 and ask what is 1 more and what is 1 less. Repeat with other numbers. Here, pupils will not have visual backup and need to work out the correct numbers mentally.
- Complete Question 1 in the Practice Book, writing the numbers in the boxes. Ask pupils which ones they did not need to count but were able to subitise.

Same-day intervention

- Prepare a bowl of ten apples and remove them one at a time, counting down until the bowl is empty. Ask pupils to tell you how many apples are in the bowl as you remove the apples.

There are five apples left in the bowl.

There are zero apples left in the bowl. The bowl is empty.

- One pupil removes some apples while others close their eyes. When they open their eyes, they must count the apples left in the bowl. Take turns.

Chapter 1 Numbers up to 10 Unit 1.8   Practice Book 1A, pages 16–17

Same-day enrichment

- Provide pupils with a selection of different types of pasta and ask them to make a pasta poster of numbers 0–10 for display. For a quick result, take a photo of the poster rather than providing glue.

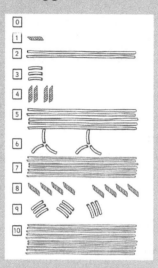

Question 2

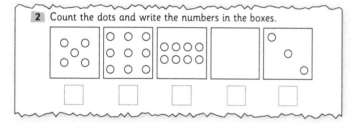

What learning will pupils have achieved at the conclusion of Question 2?

- Pupils will have further practised counting to 10 and writing numerals accurately and correctly.
- Pupils will have practised the vocabulary associated with 0: zero, nothing, nil, nought, none.
- Pupils will have continued to develop an understanding of 'same value, different appearance'.
- Pupils will have continued to develop their ability to subitise values to 4 and to recognise numbers that are arranged in standard spatial configurations.

Activities for whole-class instruction

- Discuss the vocabulary for 0.
- Ask pupils to imagine that they had eaten all the biscuits in a small packet. How would they answer their mother if she asked how many were left: *There are* **none** *left (… because I have eaten them all).*
- Ask pupils if they know how to read the results of a football match when one team does not score, for example Arsenal 2 Manchester United 0: *You say Arsenal two Manchester United* **nil**. *Nil means zero goals.*
- Show them an empty fruit bowl and ask what is in it. *There is* **nothing** *in it.* Ask how many apples are in it. *There are* **zero** *apples in the bowl.*
- Use an online 10-second countdown timer to count down with pupils to 0 seconds. Pause the timer occasionally to ask pupils what the next number will be.
- Give pairs of pupils a caterpillar diagram (or similar) with segments marked 1–10 plus red and blue counters and a 1–10 dice. One pupil rolls the dice and covers that number segment with a red counter. Their partner takes a turn, covering the number they roll with a blue counter. If the number has already been covered, no segment can be covered. The winner is the person with most counters when all the segments have counters on them.

- Give pupils opportunities to count reliably in other contexts such as:
  - sounds, for example drum beats, claps
  - movement, for example jumps or strides
  - flashes of light.

  Discuss when it may be useful to count in these contexts.

- Complete Question 2 in the Practice Book, writing the numbers in the boxes. Ask pupils which ones they did not need to count but were able to subitise.

Same-day intervention

- Mark up a blank hexagonal spinner with numbers 5–10. Pupil pairs take turns to use the spinner to generate a number and build a tower of cubes to match the number.

Same-day enrichment

- Give pupils a list of the children's names in the class. Count the number of letters in first names. Ask pupils to write and answer questions about the results. Whose name has the most letters? Whose has the least? Which number occurs most?

Chapter 1 Numbers up to 10

Unit 1.8 Practice Book 1A, pages 16–17

Question 3

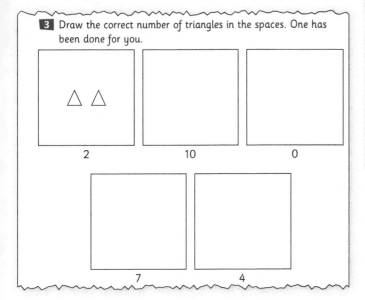

Same-day intervention

- Use a blank 10-sided dice and label it 0–9. Give pupils a pile of counters. Choose a pupil to roll the dice and ask the others to take that number of counters. Discuss with them which numbers they can 'grab' and know that they have the correct number, and which ones they need to count. Ask which is the easiest number of all, and why: *zero is the easiest because you don't need any counters at all.*

What learning will pupils have achieved at the conclusion of Question 3?

- Pupils will have further practised drawing the correct number of objects for a given number.

Same-day enrichment

- Give pairs of pupils a picture of seven cartoon cats and ask them to make up seven sentences about the picture. Each sentence must include a number.

Possible sentences for this picture are as follows.

There are seven cats in the picture. There are three black cats. There are two striped cats. There are two white cats.

One cat is standing on three paws. Four cats are standing up. Zero cats are asleep.

Activities for whole-class instruction

- Ask pupils to tell you things that there are zero of in the classroom. Remind them to say the answer in a whole sentence, for example: *There are zero elephants in the classroom. There are zero aeroplanes in the classroom.* Have fun!

- Look at fruit (or pictures of fruit) with pupils. Prepare a fruit shop front as shown for each pupil. Lay out 0–10 number cards face down and pick a card to show the number of apples in the shop that day. Fill in the number and draw the correct number of apples. Repeat for the other types of fruit.

The Juicy Fruit Shop		
apples ☐	pears ☐	plums ☐
oranges ☐	pineapples ☐	bananas ☐

- Complete Question 3 in the Practice Book, drawing the correct number of triangles in the boxes.

Challenge and extension question

Questions 4 and 5

4. Look at your ruler and find 0. What does it stand for?

5. Write a suitable number in each box.

This question introduces pupils to number lines and rulers starting from zero. The first jump is from position 0 to position 1. Look at rulers and number lines with pupils to identify zero on them.

Chapter 1 Numbers up to 10

Unit 1.9
Let's count (6)

Conceptual context

Pupils are becoming increasingly familiar with counting reliably to 10 and recognising the number of objects in standard spatial arrangements or subitising small numbers of randomly arranged objects. In this unit, they are introduced to partitioning the whole into two parts. They investigate possible sets of two numbers that make a given number up to 10. They are building their understanding that for a given number, once one of the parts is fixed, so is the other. Pupils move from a developmental stage where they 'count all' to one where they are able to 'count on' without recounting objects that they have already counted. This unit provides opportunities to 'count on'.

The part–whole concept forms the basis for knowledge of number bonds, leading to addition and subtraction calculations. Pupils are beginning to recognise the relationships between the parts and the whole, which will help them to understand formal number relationships, such as the commutative property of addition. The part–whole concept is explored further and consolidated in Chapter 2.

Learning pupils will have achieved at the end of the unit

- Pupils will be developing an understanding that there are numbers within numbers (Q1, Q2, Q3, Q4)
- Pupils will have been introduced to the whole being made of two parts (Q1, Q2, Q3, Q4)
- Pupils will be learning that for a given number, once one of the parts is fixed, so is the other (Q1, Q2, Q3, Q4)
- Pupils will have practised explaining in whole sentences how the whole can be divided into parts (Q1, Q2, Q3, Q4)
- Pupils will have been introduced to the concept of counting on (Q3)

Resources

large PE mat; counters; opaque container; double-sided counters; sorting hoop; mini whiteboards; interlocking cubes; dominoes; 1–6 dice; blank dice; large double-sided cards

Vocabulary

partition, part, whole, pattern, 1 more, 1 less, count on

Chapter 1 Numbers up to 10

Unit 1.9 Practice Book 1A, pages 18–20

Question 1

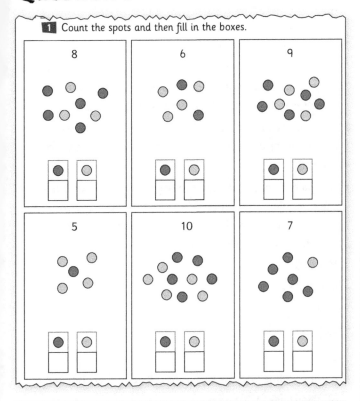

What learning will pupils have achieved at the conclusion of Question 1?

- Pupils will be developing an understanding that there are numbers within numbers.
- Pupils will have been introduced to the whole being made of two parts.
- Pupils will be learning that for a given number, once one of the parts is fixed, so is the other.
- Pupils will have practised explaining in whole sentences how the whole can be partitioned into parts.

Activities for whole-class instruction

- Choose six pupils to stand on a large PE mat. Ask: *How many pupils are standing on the mat?* Agree that there are six children on the mat.
- Roll a dice and ask that number to sit down, for example 2. Ask a pupil to tell you what they see now, for example: *There are six children on the mat. Four children are standing up and two children are sitting down.* Explain that in this case, 6 is the whole and 4 and 2 are parts that make the whole.
- Ask pupils to stand up again. Repeat rolling the dice and asking pupils to say what they see.
- Draw a large picture of a tree. Place ten small plastic birds on the picture, five in the tree and five in the air. Ask

pupils to describe what they see, for example: *There are ten birds in the picture. Five birds are in the tree. Five birds are in the air.*
- Ask a pupil to move one more bird into the tree. Describe the new arrangement. Say: *There are ten birds in the picture. Six birds are in the tree. Four birds are in the air.*
- Repeat with other instructions. Pupils can count the total number of birds to reassure themselves that the whole remains unchanged.
- Complete Question 1 in the Practice Book. Ask pupils to count the total number to check it is correct and then to count the parts that make the whole and fill in the boxes for each.

Same-day intervention

- Ask pupils to draw two pictures of an identical bookcase with two shelves (or provide a pre-drawn template). Ask them to draw seven books in each bookcase so that the two bookcases are different. Describe what they have drawn, for example: *In this bookcase, there are seven books. There are two books on the top shelf and five books on the bottom shelf.* Share their drawings.

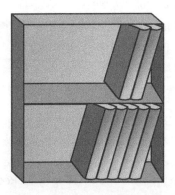

Same-day enrichment

- Count eight cubes. Place an opaque container over a small group of cubes, for example three cubes. Ask pupils if they can work out the number of cubes under the container. Listen to pupils' reasoning, for example: *I can see five cubes. I know the whole is 8. I can count on 6, 7, 8, so there are three counters under the container.*

Chapter 1 Numbers up to 10

Unit 1.9 Practice Book 1A, pages 18–20

Question 2

2. Look at the hearts in each row and then colour the dots in the same way. Write the numbers in the spaces. One has been done for you.

♥ ♡		♥	♡
♡♡♡♡♡	○○○○○		
♥♡♡♡♡	●○○○○	1	4
♥♥♡♡♡	○○○○○		
♥♥♥♡♡	○○○○○		
♥♥♥♥♡	○○○○○		
♥♥♥♥♥	○○○○○		

What learning will pupils have achieved at the conclusion of Question 2?

- Pupils will be developing an understanding that there are numbers within numbers.
- Pupils will be extending their understanding that the whole is made of two parts.
- Pupils will be learning that for a given number, once one of the parts is fixed, so is the other.
- Pupils will have practised explaining, in whole sentences, how the whole can be divided into parts.

Activities for whole-class instruction

- Place six double-sided black and white counters in a pot and ask a pupil to tip them into a sorting hoop. Record together on a table as shown how many black and how many white faces there are. Put the counters back into the pot, give it a good shake and tip them out again. Repeat about eight times. Some arrangements will occur more than once.

All about 6		
	black counters	white counters
1.	●●●●	○○
2.	●●●	○○○
3.	●	○○○○○
4.	●●●●	○○
5.	●●●●●	○
6.	●●●	○○○
7.	●●	○○○○
8.		○○○○○○

Ask pupils to tell you, in a whole sentence, if any of the throws are the same, for example: *Yes, number 1 and number 4 are they same. They both have four black counters and two white ones.*

Ask if they think there are any other possible arrangements. *Number 8 has six white counters but there isn't one where all the counters are black. 'Six black counters' is missing.*

- Give pupils, in pairs, a blank recording sheet and ask them to do the same investigation with another number, for example 4 or 7.
- Ask pairs of pupils if they can use counters to set out all the possible arrangements for four double-sided counters in a systematic way. Help them to establish the pattern and share the results.

●●●●

●●●○

●●○○

●○○○

○○○○

Point to a line and ask pupils to tell you what they see, for example:

●●●○

There are four counters. Three counters are black and one counter is white.

- Complete Question 2 in the Practice Book, colouring the dots and writing down the numbers. Ask pupils to describe what they see and explain the patterns.

Same-day intervention

- In pairs, pupils set out a random pattern of seven cubes in a sorting hoop. They take turns to place a piece of string (or a ruler) across the hoop so that some cubes are on one side and some on the other. Describe the whole and parts in a complete sentence.

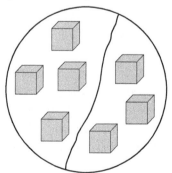

There are seven cubes. Four are on one side and three on the other.

Repeat with other numbers.

Chapter 1 Numbers up to 10

Unit 1.9 Practice Book 1A, pages 18–20

Same-day enrichment

- Look at the patterns in the question. Challenge pupils to draw up a similar table for another number. They can choose how many hearts to start with (7, 8, 9 or 10).

Question 3

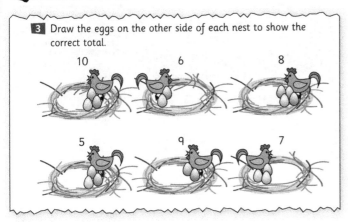

3 Draw the eggs on the other side of each nest to show the correct total.

What learning will pupils have achieved at the conclusion of Question 3?

- Pupils will be developing an understanding that there are numbers within numbers.
- Pupils will be extending their understanding that the whole is made of two parts.
- Pupils will be learning that, for a given number, once one of the parts is fixed, so is the other.
- Pupils will have practised explaining in whole sentences how the whole can be divided into parts.
- Pupils will have been introduced to the concept of counting on.

Activities for whole-class instruction

- Ask pupils to imagine seven children. Think about how many combinations of boys and girls are possible. Try a few examples, counting out one group of boys or girls. Discuss how you can work out the second part once the first one has been determined. Practise counting on. For example: *The whole is seven children. There are five boys. Starting at 5, I count on 6, 7, so there are two girls.*
- In pairs, invite pupils to make 10. Roll a 1–6 dice to determine the number of objects in one part and then challenge them to work out the number required for the other part. Ask them to record their results by showing the two parts within a whole sorting ring. Help them to count on to find the second part.
- Complete Question 3 in the Practice Book, drawing the correct number of eggs to show the correct total.

Same-day intervention

- Make 7. Roll a dice marked 0–5 to determine how many objects should be in one part. Ask pupils to work out how many objects should be in the other side.

Same-day enrichment

- Investigate how the dots are arranged on a dice. Ask which numbers are opposite one another on their dice and check that this arrangement is the same on all dice. Ask pupils to explain the pattern to a talk partner. Roll a dice and ask them to predict the hidden number. Check that they are right!

Question 4

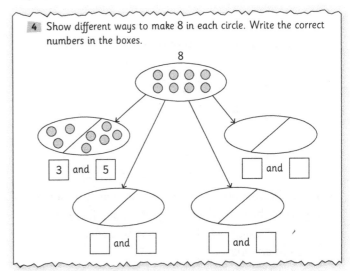

4 Show different ways to make 8 in each circle. Write the correct numbers in the boxes.

What learning will pupils have achieved at the conclusion of Question 4?

- Pupils will be developing an understanding that there are numbers within numbers.
- Pupils will be extending their understanding that the whole is made of two parts.
- Pupils will be learning that for a given number, once one of the parts is fixed, so is the other.
- Pupils will have practised explaining in whole sentences how the whole can be divided into parts.

Chapter 1 Numbers up to 10

Unit 1.9 Practice Book 1A, pages 18–20

Activities for whole-class instruction

- Choose four pupils and ask them to sit in a line in front of the rest of the class. Count them and establish that there are four. Give each pupil a large double-sided card, for example blue on one side and yellow on the other. Ask how many cards there are. Establish that there are four cards because each pupil has only one card. Explain that you are going to count down together and when you get to zero, they should hold up one side of the card. Count down from 4. Ask the class to describe what they can see in whole sentences. For example: *There are four cards. Three are yellow. One is blue. Three and one is four.*

  Tell pupils you are going to repeat it and they can keep their card the same colour or change it. Repeat as appropriate. Discuss the results.

- Increase the number of pupils to nine and repeat the exercise, this time counting down from 9.

- Complete Question 4 in the Practice Book, drawing a correct numbers of dots on either side of the oval to make 8.

Same-day intervention

- Show pupils the set of dominoes that total six spots. Ask them to count the spots to check that they all have six. Turn them over and mix them up.

- Ask individual pupils to choose a domino, turn it over and describe what they see, for example: *There are six spots on the domino. One part has five spots; the other part has one spot.*

- Use a full set of dominoes (remove those with totals of 11 and 12) and repeat. Now pupils first need to count the total number of spots and then work out the spots on each side.

Same-day enrichment

- Ask pupils to look at their answer to Question 4 and discuss if they have found all the possibilities. (There is one missing possibility, 0 and 8 / 8 and 0.) This might lead to an early discussion of the commutative aspect of addition.

Challenge and extension question

Question 5

5 Write numbers on the boxes to complete the pattern.

This question introduces pupils to doubles. Sufficient information has been provided to show the growing pattern and they have already investigated 5 and 5 is 10. To explore this further, pupils could look for the doubles in a set of dominoes.

Chapter 1 Numbers up to 10

Unit 1.10
Counting and ordering numbers (1)

Conceptual context

In this unit, pupils are introduced to ordinal numbers: first, second, third ... Ordinal numbers define the position of objects in a series. Position 1 is first, position 5 is fifth (5th); thus the cardinal numerals match the ordinal ones. Pupils may be familiar with first, second and third from competitions or sports results.

When objects are arranged horizontally, by convention they should be counted and ordered from the left. However, this can be overruled if an instruction begins, 'Starting from the right'.

(i) An ordinal number is a number that defines position in a list. Most ordinal numbers end in 'th', except for 1st (fi**rst**), 2nd (seco**nd**), 3rd (thi**rd**).

Look out for ... for pupils who find it difficult to identify left and right quickly and consistently. Help them to develop a technique to determine left and right, such as such as placing their thumbs at right angles to their index fingers. The hand that makes an 'L' is their left hand.

Check too for pupils who have difficulty pronouncing ordinal numbers correctly. The pronunciation is tricky and worth specific practice.

Cardinal	Ordinal	
1	1st	first
2	2nd	second
3	3rd	third
4	4th	fourth
5	5th	fifth
6	6th	sixth
7	7th	seventh
8	8th	eighth
9	9th	ninth
10	10th	tenth

Learning pupils will have achieved at the end of the unit

- Pupils will have practised counting and ordering numbers to 10 (Q1, Q2, Q3)
- Pupils will have practised using the vocabulary of ordinal numbers (Q1, Q2, Q3)
- Pupils will have followed instructions concerning ordinal numbers and explored the importance of reading questions very carefully (Q3)

Resources

small world objects (e.g. animals, vehicles); 1–10 number cards (individual and class); ordinal number word cards (first, second, third, fourth) (individual and class); counters; interlocking cubes

Vocabulary

ordinal numbers, first, second, third, fourth, fifth, sixth, seventh, eighth, ninth, tenth, left, right

Chapter 1 Numbers up to 10

Unit 1.10 Practice Book 1A, pages 21–23

Question 1

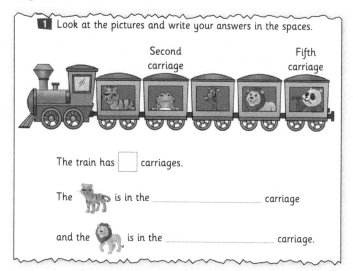

What learning will pupils have achieved at the conclusion of Question 1?

- Pupils will have practised counting and ordering numbers to 5.
- Pupils will have used the vocabulary of the first 5 ordinal numbers.

Activities for whole-class instruction

- Use chairs to make a 'train' in the classroom with five chairs as 'carriages'. Use ordinal cards from 'first' to 'fifth' and discuss how to place them correctly on the chairs.
- Ask individual pupils to sit in specific places, say: *Please sit in the third carriage.*
  Remove the labels and give pupils similar instructions and ask them to count to find the correct place, say: *Please sit in the fourth carriage.* Pupil counts: *First, second, third, fourth. This is the fourth carriage.*
- Set out five coloured counters. Explain that they represent stepping stones across a river from an island to home.

	R	I	V	E	R	
	colour a	colour b	colour c	colour d	colour e	
ISLAND	●	●	●	●	●	HOME
	R	I	V	E	R	

- Ask pupils questions that use ordinal numbers:
  What colour is the second stepping-stone?
  What number stepping-stone is the green one?
  Continue with further questions, ensuring that pupils answer in whole sentences.
- Complete Question 1 in the Practice Book, filling in the boxes.

Same-day intervention

- Shuffle five cards with the first ordinal numbers on them: first, second, third, fourth and fifth. Give them to five pupils and ask them to order themselves in a line correctly. Ask each pupil to describe their position in a complete sentence, for example: *I am in third place, second place is in front of me and fourth place is behind me.*

  Now ask pupils to turn sideways and discuss what they will look like now. They still have the same positions, for example: *I am in third place; second place is on one side of me and fourth place on the other side.*

Same-day enrichment

- Show pupils a picture of a cartoon bus. Count the windows and ask whom they will see in the first window. Establish that it is the driver. Ask them to draw the driver and some passengers.

  Using ordinal numbers, ask them to describe in whole sentences the places on the bus:

  The driver is in the first seat.
  The fifth seat is empty.
  The lady in the fourth seat has curly hair.
  The man in the sixth seat is wearing a hat.

  Keep reminding pupils to start counting from the driver.

Chapter 1 Numbers up to 10

Unit 1.10 Practice Book 1A, pages 21–23

Question 2

> **2** Count and then write your answers in the spaces.
>
> (a)
>
> left   right
>
> There are ☐ pieces of fruit altogether.
>
> Counting from the left, 🍉 is in the _____ position, and 🍍 is in the _____ position.
>
> Counting from the right, 🍉 is in the _____ position, and 🍍 is the _____ position.
>
> (b)
>
> Counting from the left, 🐱 is in the _____ position.
>
> Counting from the right, 🐱 is in the _____ position.
>
> There are ☐ animals in total.
>
> 🐼 is in the middle.
>
> There are ☐ animals on its left.
>
> There are ☐ animals on its right.
>
> There are ☐ animals on the left of 🦋.

What learning will pupils have achieved at the conclusion of Question 2?

- Pupils will have practised counting and ordering numbers to 10.
- Pupils will have practised using the vocabulary of ordinal numbers.

Activities for whole-class instruction

- Use two sets of class cards, 1–10 number cards and first to tenth word cards. Shuffle them and give them out to pupils. Ask who has 1 and what position that is. Establish that 1 is 1st/first. Continue to build up to the tenth place.

1	first	6	sixth
2	second	7	seventh
3	third	8	eighth
4	fourth	9	ninth
5	fifth	10	tenth

- Ask ten pupils to line up, facing the person in front as if they were leaving the classroom. Ask pupils questions that use ordinal number, say: *Who is first in the line? Who is in third place? What is the position of the pupil who is last in line?*
- Ask pupils to turn so that they are standing side by side, still in the same order. Which pupils are standing next to the pupil who is seventh?

Challenge pairs of pupils to ask each other and answer more questions using ordinal numbers. Share some of their questions.

- Set out ten vehicles (or pictures of vehicles) in a straight line.

| bike | van | lorry | car | scooter | tractor | sports car | taxi | bus | motorcycle |

Check that pupils know their left and right hands and ask them to point to the left side of the line and name the vehicles in order. Count the vehicles.

Ask pupils questions that use ordinal numbers:

How many vehicles are there in total? (10)
Counting from the left, what vehicle is in fourth place? (car)
Counting from the left, what position is the taxi in? (eighth)
Counting from the right, what position is the taxi in? (third)
Counting from the right, what vehicle is in sixth place? (scooter)

Continue asking questions to check understanding.

- Complete Question 2 in the Practice Book, counting and filling in the boxes.

Same-day intervention

- Ask pupils to look at the line of vehicles used in the whole-class activity. Ask them what position the tractor is in:
  - starting from the left
  - starting from the right.

  Repeat with other vehicles.

- There are many online games available to give pupils additional practice.

Same-day enrichment

- Working in groups of about five, challenge pupils to arrange themselves in alphabetical order of their first names. Record the results.

first	
second	
third	
fourth	
fifth	

Chapter 1 Numbers up to 10

Unit 1.10 Practice Book 1A, pages 21–23

Question 3

3 Count and colour the hearts. Start from the left.
Colour five of the hearts:

Colour the fifth heart:

What learning will pupils have achieved at the conclusion of Question 3?

- Pupils will have practised counting and ordering numbers to 10.
- The connection between ordinal and cardinal numbers will have been discovered.

Activities for whole-class instruction

- Draw ten circles in a horizontal line. Read pupils these instructions and work together to colour the circles so that they match the description.

  Starting from the left the first circle is yellow.

  The third and fifth circles are red.

  The seventh and eighth circles are blue.

  The tenth circle is green.

  The remaining circles are purple.

- Give pupils mini whiteboards and ask them to draw a line of seven triangles.

  Start from the left and ask them to colour the first three triangles blue. Then to colour the sixth triangle red. Ask how many triangles have not been coloured. *Three triangles have not been coloured.* Which ones are they? *The fourth, fifth and seventh have not been coloured.*

  Choose individual pupils to give further instructions about colouring triangles until they are all coloured.

- Complete Question 3 in the Practice Book, colouring in the hearts.

Same-day intervention

- Show pupils two rows of five coloured cubes arranged as below.

top	Y	Y	Y	B	G
bottom	R	Y	B	G	R

- Ask them which line (top, bottom or both) fits each statement. They should start counting from the left and give their response in a full sentence.
  - The first cube is red. *The first cube is red in the bottom row.*
  - There are three yellow cubes. *There are three yellow cubes fits the top row.*
  - The fifth cube is green. *The fifth cube is green fits the top row.*
  - The second cube is yellow. *The second cube is yellow fits both rows.*

Same-day enrichment

- Complete the same-day intervention.
  Challenge pupils to design their own two rows of five cubes, and write statements and answers. Remember to count from the left!

Challenge and extension question

Question 4

4 Draw the missing shapes in the boxes.

(a) Starting from left to right, ▲ is in the fourth place. How many △ are there on its left?

▲ △ △ △ △

(b) Starting from right to left, ● is in the third place. How many ○ are there on its right?

○ ○ ○ ●

In this question, pupils read and carry out instructions involving ordinal numbers, counting from both left and right.

Chapter 1 Numbers up to 10

Unit 1.11
Counting and ordering numbers (2)

Conceptual context

In the previous unit, pupils were introduced to ordinal numbers: first, second, third … ; numbers that define the position of objects in a series. In this unit, understanding of ordinal numbers is consolidated and pupils will understand and write ordinal numbers as words.

ⓘ Cardinal numbers are **c**ounting numbers, answering the question 'How many?'
Nominal numbers are numbers used as **n**ames, for example the number on a footballer's shirt or on a bus.

Ordinal numbers tell the **o**rder of things in lists.

When ordinal numbers are expressed as figures, the last two letters of the written word are added to the cardinal number.

fir**st**	1st
seco**nd**	2nd
thi**rd**	3rd
four**th**	4th
fif**th**	5th
six**th**	6th
seven**th**	7th
eigh**th**	8th
nin**th**	9th
ten**th**	10th

Learning pupils will have achieved at the end of the unit

- Pupils will have consolidated their understanding of counting and ordering numbers to 10 (Q1, Q4)
- Pupils will have practised describing position using ordinal numbers (Q1, Q2, Q3, Q4)
- Pupils will have practised reading and writing ordinal numbers as numbers and words (Q1, Q2, Q3, Q4)
- Pupils will have revised carefully following instructions concerning ordinal numbers (Q1, Q2, Q3, Q4)
- Pupils will have ordered objects by different criteria, for example length, height (Q2, Q3)

Resources

small world objects (e.g. animals, vehicles); 1–10 number cards (individual and class); ordinal number cards as numbers (1st, 2nd, 3rd, 4th …) and as words (first, second, third, fourth) (individual and class); blank cards; mini whiteboards; bricks; interlocking cubes; foam javelins; blank dice

Vocabulary

1st, 2nd, 3rd, 4th, 5th, 6th, 7th, 8th, 9th, 10th, first, second, third, fourth, fifth, sixth, seventh, eighth, ninth, tenth

Chapter 1 Numbers up to 10

Unit 1.11 Practice Book 1A, pages 24–25

Question 1

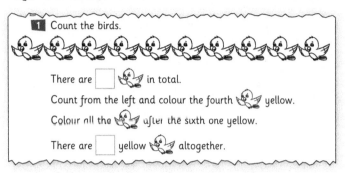

What learning will pupils have achieved at the conclusion of Question 1?

- Pupils will have consolidated their understanding of counting and ordering numbers to 10.
- Pupils will have practised describing position using ordinal numbers.
- Pupils will have practised reading and writing ordinal numbers as numbers and words.
- Pupils will have revised carefully following instructions concerning ordinal numbers.

Activities for whole-class instruction

- Use a set of 1st–10th cards for pupils to generate a pattern. Shuffle the cards and lay them face down in front of pupils.

  Give pupils mini whiteboards and ask them to draw a row of ten squares. Ask a pupil to choose a colour and then turn over a card. Counting from the left, they should colour that square on their board, for example: *I chose blue. This card says 3rd. I will colour the third square blue.*

  Continue choosing other pupils until all the cards have been turned over. Show and compare patterns on the mini whiteboards.

- Show pupils a line of animals and ask them to complete statements similar to those described.

  There are _____ animals in total.
  There are _____ different types of animal.
  Count from the left each time.
  The cows are in _____ , _____ and _____ places.
  The chickens are in _____ and _____ places.
  The animals in the seventh and ninth places are _____ .

The animals in the fourth and tenth places are _____ .

- Complete Question 1 in the Practice Book, colouring the birds according to the instructions.

Same-day intervention

- Discuss with pupils how racing cars begin a race by lining up in order according to their times in practice. Draw a row of ten (simple) racing cars facing right on the board. Elicit from pupils that the first car is the one on the right and emphasise this. Count from first to tenth together. Give individual pupils instructions to complete the grid, working through this together. Counting from the right:

  - A blue car is in first place.
  - The fourth car is yellow.
  - A red car is in fifth place.
  - A purple car is in tenth place.
  - The seventh car is black.

  - A pink car is in ninth place.
  - The orange car is in second place.
  - A green car is in sixth place.
  - A brown car is in third place.
  - The grey car is in eighth place.

- When the picture is complete, ask pupils to answer and make up questions.

Same-day enrichment

- Ask pupils to set out ten pieces of fruit in a line and make up five statements using the following words. Different words can be chosen; these are the trickiest ones to read and spell.

| first | fourth | eighth | third | fifth |

Chapter 1 Numbers up to 10

Unit 1.11 Practice Book 1A, pages 24–25

Question 2

What learning will pupils have achieved at the conclusion of Question 2?

- Pupils will have ordered objects by different criteria, for example length, height.
- Pupils will have practised describing position using ordinal numbers.
- Pupils will have practised reading and writing ordinal numbers as numbers and words.
- Pupils will have revised carefully following instructions concerning ordinal numbers.

Activities for whole-class instruction

- Ask for five volunteers to jump from a standing start and mark (do not measure) the distance jumped. Give cards marked with ordinal numbers to indicate who was first, second and so on. Choose individual pupils to describe the position of their peer in a full sentence, for example: *Sam is in first place because he jumped the furthest.*
- Build brick towers of 5, 7, 4, 10 and 3 cubes.
- Ask pupils to tell you which is the tallest tower, and then the next tallest, down to the shortest.
- Give pupils mini whiteboards and ask them to draw a table like this, marking the position of the tallest tower as 1st. Continue with the rest of the towers. Share whiteboards.

Tower of 5 bricks	Tower of 7 bricks	Tower of 4 bricks	Tower of 10 bricks	Tower of 3 bricks
			1st	

- Complete Question 2 in the Practice Book, putting the heights in order.

Same-day intervention

- A wildlife photographer in Africa keeps a list of the number of leopards he spots each day. Use interconnecting cubes to make towers matching the number of cubes to the number of leopards he saw each day. Help pupils to order the days from first to seventh, ranking the number of leopards he saw, starting with the day he saw most leopards.

Number of leopards seen each day			
Sunday	Monday	Tuesday	Wednesday
3	8	10	6
Thursday	Friday	Saturday	
5	0	8	

Same-day enrichment

- Pupils put eight snakes in order of length from first to eighth, beginning with the longest one.

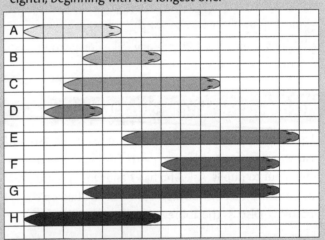

Snakes in order of length:

1st	
2nd	
3rd	
4th	
5th	
6th	
7th	
8th	

Chapter 1 Numbers up to 10   Unit 1.11 Practice Book 1A, pages 24–25

Question 3

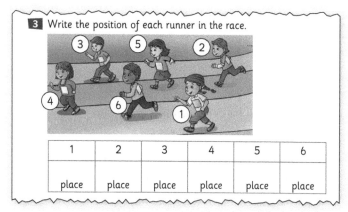

3 Write the position of each runner in the race.

1	2	3	4	5	6
place	place	place	place	place	place

- If appropriate, give the words, first to tenth, as a spelling exercise in a Literacy lesson.
- Complete Question 3 in the Practice Book, filling in the boxes. Point out to pupils that the numbers circled next to the children are numbers being used simply to identify them.

Same-day Intervention
- Prepare cards showing the top ten pets in the UK

1st dogs	2nd cats	3rd fish	4th rabbits	5th hamsters
6th lizards	7th caged birds	8th guinea pigs	9th chickens	10th snakes

- Shuffle the cards and lay them out randomly. Choose a pupil to find the first one and then more pupils to place them in order. Discuss the top ten. Are there any surprises? What pets do they have? Ask each pupil to describe one of the cards using an ordinal number. *The dog is in first place. Lizards are in sixth place – that is surprising because I don't know anyone with a lizard.*

What learning will pupils have achieved at the conclusion of Question 3?

- Pupils will have ordered objects by different criteria, for example length, height.
- Pupils will have practised describing position using ordinal numbers.
- Pupils will have practised reading and writing ordinal numbers as numbers and words.
- Pupils will have revised carefully following instructions concerning ordinal numbers.

Same-day enrichment
- Show pairs of pupils this list of top ten tunes for pre-school children (or a similar 'top ten' list):
  1st Humpty Dumpty
  2nd Row, row, row your boat
  3rd Ring a ring a roses
  4th Twinkle twinkle little star
  5th The wheels on the bus
  6th Incey wincey spider
  7th Old MacDonald
  8th Five little monkeys jumping on the bed
  9th This is the way the lady rides
  10th Five little ducks went swimming one day
- Ask pupils whether they know all these songs and if they agree with the order. If not, discuss the order in which they would place them.
- Challenge them to write their own top ten, for example for nursery rhymes, TV programmes and so on.

Activities for whole-class instruction

- Use three sets of class cards, 1–10 number cards, 1st–10th cards and first to tenth word cards. Shuffle them and give them out to pupils. Ask who has 1 and what position that is. Establish that 1 is 1st/first. Continue to build up the patterns. Discuss the abbreviations with pupils.

1	1st	fi**rst**
2	2nd	seco**nd**
3	3rd	thi**rd**
4	4th	four**th**
10	10th	ten**th**

- Label ten foam PE javelins, A–J. Ask for volunteers or choose pupils to throw the javelins. Once all the javelins have been thrown, complete a table as shown with the identifying letters to record the order travelled.

1st	2nd	3rd	4th	5th	6th	7th	8th	9th	10th

Question 4

4 Draw shapes according to the instructions.

Counting from the left, draw one △ in the 5th place, one ○ in the 10th place, one ☐ in the 3rd place, and two ♡ in the 7th place.

Chapter 1 Numbers up to 10

Unit 1.11 Practice Book 1A, pages 24–25

What learning will pupils have achieved at the conclusion of Question 4?

- Pupils will have consolidated their understanding of counting and ordering numbers to 10.
- Pupils will have practised describing position using ordinal numbers.
- Pupils will have practised reading and writing ordinal numbers as numbers and words.
- Pupils will have revised carefully following instructions concerning ordinal number.

Activities for whole-class instruction

- Prepare cards with instructions (words or pictures) such as:
  - Sit down
  - Clap your hands quietly
  - Tap your foot on the floor
  - Put your hands on your head
  - Touch your nose
  - Fold your arms
  - Put your thumbs up
  - Kneel down
  - Wiggle your fingers
  - Put your hands on your hips

  Ask ten pupils to stand in a line and to describe their position starting from the left. For example: *I am first in line; I am second from the left; I am third.* Place a card in front of each pupil and ask them to carry out the instruction.

  Ask other pupils to describe what a particular child is doing. Ask: *What is the fourth child from the left doing? What is the third child from the right doing?*

  Repeat with different pupils and reshuffled cards.

- Ask pupil pairs to build a line of interlocking cubes. Mark up a blank dice with three faces marked 'left' and three face marked 'right'.

  Pupils take turns to roll the dice to decide whether to count from the left or right and then to ask their partner a question, such as:

  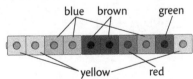

  Counting from the left, what colour is the seventh cube?
  Counting from the left, in what positions are the yellow cubes?
  Counting from the right, what colour cube is in fifth place?
  Counting from the right, in what position is the blue cube?

- Complete Question 4 in the Practice Book, drawing figures according to the instructions.

Same-day intervention

- Here are the top ten most dangerous wild animals in the world in reverse order.

10. hippopotamus		5. rhinocerous
9. leopard		4. brown bear
8. cheetah		3. hyena
7. polar bear		2. buffalo
6. elephant		1. lion

  Ask: *Which is the most dangerous? The second most dangerous?* Give this list to pupils and ask them to order it from first to tenth. Ask each pupil to make up a question for their peers to answer.

Same-day enrichment

- Draw six circles in a line. Counting from the right, draw:
  - 10 dots in the fourth circle
  - 5 dots in the second circle
  - 1 dot in the first circle
  - 8 dots in the third circle
  - 6 dots in the fifth circle
  - 3 dots in the sixth circle.

  Now write ordinal numbers below the circles, to show the circle with most dots as first, second the next highest number of dots, down to sixth as the circle with the fewest dots.

Challenge and extension question

Question 5

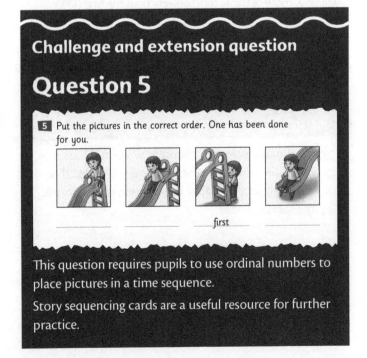

This question requires pupils to use ordinal numbers to place pictures in a time sequence.

Story sequencing cards are a useful resource for further practice.

Chapter 1 Numbers up to 10

Unit 1.12
Let's compare (1)

Conceptual context

In this unit, pupils are learning to count and compare sets of objects to determine which set has more or fewer objects. Pupils use one-to-one correspondence to see visually the set with more objects. In some questions they will compare two sets and in others, three sets. As they build experience of the size of numbers, the number words themselves become countable and comparable items. This will be explored in the next unit.

Learning pupils will have achieved at the end of the unit

- Pupils will have compared two sets of objects by using one-to-one correspondence (Q1, Q3, Q4)
- Pupils will have compared three sets of objects by using one-to-one correspondence (Q2)
- Pupils will have learned to identify the set that has more/most objects (Q1, Q2, Q3, Q4)
- Pupils will have practised describing sets in whole sentences to say which has more/most objects (Q1, Q2)
- Pupils will have learned to count, compare and identify the set that has more, fewer or the same number of objects (Q3, Q4)
- Pupils will have recognised that if there are X fewer of the first objects than the second, then there will be X more of the second objects than the first (Q3)
- Pupils will have learned to describe sets of objects in full sentences using more, fewer or the same (Q3, Q4)
- Pupils will have learned to draw sets of objects, following instruction sentences that use the words more, fewer or the same (Q4)

Resources

mini whiteboards; coloured counters; white and blue dice; bags of marbles; plates; envelopes; stamps; 10-sided dice marked with digits 0–9; paper cups; plastic cutlery; blank dice ; interlocking cubes

Vocabulary

compare, sets, more, most, fewer, same

Chapter 1 Numbers up to 10

Unit 1.12 Practice Book 1A, pages 26–28

Question 1

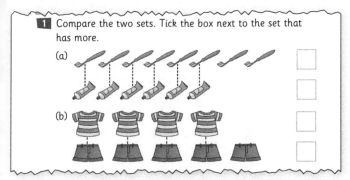

What learning will pupils have achieved at the conclusion of Question 1?

- Pupils will have compared two sets of objects by using one-to-one correspondence.
- Pupils will have learned to identify the set that has more objects.
- Pupils will have practised describing sets in whole sentences to say which has more objects.

Activities for whole-class instruction

- Roll a white dice and count out that number of white counters. Lay them in a line.

  Roll a blue dice and count out that number of blue counters, laying them in a line to match the white ones. Ask pupils to compare the two sets to demonstrate which set has more, for example: *There are five white counters and three blue counters. There are more white counters.*

  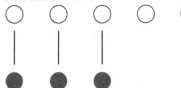

  Ask: *Do you need to actually count the counters to see which row has the most?*

- Lay out two rows of counters in line with each other and ask: *Which row has the most counters – can you tell me without counting?*

- Bring two pupils to the front and give each a bag of marbles and a plate, each bag containing a different number of marbles (numbers not known to pupils). Ask the class: *How can we find out which bag has the most marbles?* Gather ideas. Ask the two pupils to each take a marble from their bag and put it on a plate. They should repeat, in time with each other, until one pupil has no marbles left in their bag. Ask: *What do we know about which bag has the most marbles?*

- Complete Question 1 in the Practice Book, putting a tick against the set that has more.

Same-day intervention

- Prepare three trays of envelopes and stamps.
  - Tray 1: four envelopes and five stamps
  - Tray 2: seven envelopes and six stamps
  - Tray 3: nine envelopes and ten stamps
- Ask pupils to count the envelopes and stamps tray by tray, match them and explain whether there are enough stamps for all the envelopes.

  For example: *Tray 1 has more stamps than envelopes so there are enough stamps for all the envelopes.*
- Repeat for the other trays.

Same-day enrichment

- Write the numbers 1 to 10. In pairs, pupils take turns to roll a 10-sided dice marked with digits 0–9. The first pupil covers the number that is 1 more with a red counter, for example roll a 4, cover 5. The second player uses blue counters. The winner has the most counters.

Question 2

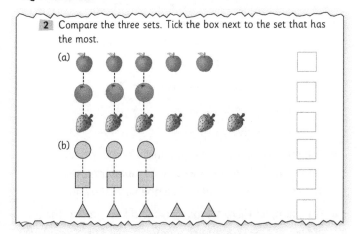

What learning will pupils have achieved at the conclusion of Question 2?

- Pupils will have compared three sets of objects by using one-to-one correspondence.
- The word 'most' will be understood in this context.
- Pupils will have practised describing sets in whole sentences to say which has most objects.

Chapter 1 Numbers up to 10

Unit 1.12 Practice Book 1A, pages 26–28

Activities for whole-class instruction

- Prepare ten paper cups with the same coloured counters in them as follows:
  - 2 cups with 6 counters in each
  - 2 cups with 7 counters in each
  - 2 cups with 8 counters in each
  - 2 cups with 9 counters in each
  - 2 cups with 5 counters in each.
- Arrange the cups randomly. Invite three pupils to choose a cup. Tip out the counters in the first cup, arrange them in a line and count them. Repeat with the other two cups, matching the counters. Ask the class which line has the most counters.

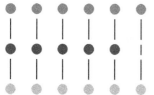

The third set has the most.

- Repeat with three more cups.
- Complete Question 2 in the Practice Book, comparing three sets and ticking the set that has the most.

Same-day intervention

- Prepare four trays of plastic knives, forks and spoons where one set has equal numbers, for example:
  - 4 forks, 5 knives, 4 spoons
  - 6 forks, 8 knives, 7 spoons
  - 5 forks, 5 knives, 5 spoons
  - 7 forks, 6 knives, 5 spoons.
- Ask pupils to choose one tray and arrange the cutlery to determine if there are the same number or more forks, knives and spoons, and to describe what they find.
- For example: *There are four forks and four spoons. There are five knives. I can match four forks, four knives and four spoons. There is one knife that is not matched. There is one more knife. There are more knives than forks and spoons.*
- Repeat with the other trays.

Same-day enrichment

- Show pupils a random mixture of three different coloured cubes (or a picture of cubes) and ask pupils to count the cubes and make a statement describing which colour has most cubes. For example: *There are 8 red cubes, 10 green cubes and 9 blue cubes. There are most green cubes.*

Question 3

3 Compare the sets and write the missing numbers.

There are ☐ 🧊. There are ☐ 🟥.
There are ☐ fewer 🧊 than 🟥.
There are ☐ more 🟥 than 🧊.

There are ☐ ☆. There are ☐ ★.
There are ☐ fewer ☆ than ★.
There are ☐ more ★ than ☆.

Chapter 1 Numbers up to 10

Unit 1.12 Practice Book 1A, pages 26–28

What learning will pupils have achieved at the conclusion of Question 3?

- Pupils will have learned to count, compare and identify the set that has more, fewer or the same number of objects.
- Pupils will have recognised that if there are X fewer of the first objects than the second, then there will be X more of the second objects than the first.
- Pupils will have learned to compare sets of objects in full sentences using more, fewer or the same.

Activities for whole-class instruction

- Show pupils three circles and five squares.
  Ask if there are more circles or squares. *There are two more squares than circles.*
  Ask if there are fewer circles or squares. Help pupils to describe that there are two **fewer** circles than squares. This may be a new word for them.
  Ask them to complete the sentence: *There are _____ circles than squares.*
  Repeat with other numbers of circles and squares to establish that if there are X fewer circles than squares there will be X more squares than circles.
- In pairs, using mini whiteboards, ask pupils to roll a 1–6 dice and draw that number of circles. Roll a second time and draw that number of squares. (Roll again if the numbers are the same.) Use the words 'more' and 'fewer' to describe the results.
  Roll 2 ●●   Roll 4 ☐☐☐☐
  There are 2 fewer circles than squares.
  There are 2 more squares than circles.
- Complete Question 3 in the Practice Book, filling in the blanks.

Same-day intervention

- Ask pupils to draw circles to complete the grid, showing one fewer and one more than the central number. The first line has been completed.

1 fewer		1 more
●●●●●●	7	●●●●●●●●
	3	
	10	
	6	
	9	

Same-day enrichment

- Give pairs of pupils a copy of this story.
  One day, a family went for a walk in the woods, looking for wildlife. They hadn't been walking long before they spotted a group of deer. Startled, the six deer quickly bounded away. Further on in a clearing in the woods they spotted nine rabbits, nibbling the grass. The family stood very still; the rabbits looked up but did not run away.
- Pupils draw simple diagrams of the animals and match them. Then they answer these questions:
  - Were there fewer rabbits or deer? *There were three fewer deer than rabbits.*
  - Were there more rabbits or deer? *There were three more rabbits than deer.*

Question 4

4 Compare and then draw.

On the first line, draw 4 ◯.

On the second line, draw △, so there are 3 more than ◯.

On the third line, draw ☐, so there are 2 fewer than ◯.

What learning will pupils have achieved at the conclusion of Question 4?

- Pupils will have learned to draw sets of objects, following instruction sentences that use the words more, fewer or the same.
- Pupils will have learned to describe sets of objects in full sentences using more, fewer or the same.

Activities for whole-class instruction

- Take a blank dice; mark the faces '1 more', '1 fewer', '2 more', '2 fewer', '3 more' and '3 fewer'.
- Separate interlocking cubes into different colours.
  Ask a pupil to count eight red cubes and place them in a line.
- Choose another pupil to roll the dice and carry out the instruction using blue cubes, for example roll '2 fewer', count out and match six blue cubes, placing them above

Chapter 1 Numbers up to 10

Unit 1.12 Practice Book 1A, pages 26–28

the red cubes. Choose a third pupil to make a third line using yellow cubes, for example roll '1 more', count out and match nine yellow cubes, placing them below the red cubes.

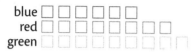

- Ask questions about the cubes, for example compare the blue and red lines. For example: *There are 2 more red cubes than blue cubes. There are 2 fewer blue cubes than red cubes.*
- Repeat, starting with a different number of red cubes.
- Complete Question 4 in the Practice Book, drawing the correct number of shapes.

Same-day intervention

- Say to pupils: *In the classroom there is a selection of fruit on a tray. There are five apples. There is 1 more orange than apples. There are 2 fewer bananas than apples. By comparing, draw a diagram to illustrate the number of each type of fruit.*

Same-day enrichment

- Ask pupils to carry out the Same-day intervention task. Now make up and write instructions for a tray of pears, plums and peaches.

Challenge and extension question

Question 5

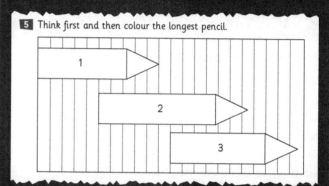

5 Think first and then colour the longest pencil.

In this question, pupils need to think about the length of each pencil, from base to point. This can be achieved by using the background columns and requires counting beyond 10.

Chapter 1 Numbers up to 10

Unit 1.13
Let's compare (2)

Conceptual context

The symbols > (greater than), = (equal to) and < (less than) are introduced in this unit to enable sets of objects to be compared using a number expression.

| 6 < 8 | 8 = 8 | 10 > 8 |

Unlike adults, pupils do not automatically know if a number is greater than or less than another number. They need plenty of practical experience with manipulatives and numbers to be able to visualise these relationships mentally. Encourage them to think about the relative sizes of numbers.

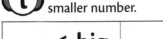 The symbols > and < are easily confused. Remember the 'small' end always points to the smaller number.

| small < big | big > small |

When two values are the same, we say they are equal and use the = symbol, for example 5 = 5.

Learning pupils will have achieved at the end of the unit

- Pupils will have compared numbers to 10 and identified the bigger/smaller number (Q1, Q2, Q3, Q4, Q5)
- Pupils will have practised completing number statements using the >, = and < symbols (Q1, Q2, Q3, Q4)
- Pupils will have practised comparing numbers mentally and using >, < and = to write number statements (Q4, Q5)

Resources

0–10 number cards (individual and class); >, < and = cards (individual and class); 1–10 spinners; mini whiteboards

Vocabulary

greater than (>), less than (<), same as/equal to (=)

Chapter 1 Numbers up to 10

Unit 1.13 Practice Book 1A, pages 29–30

Question 1

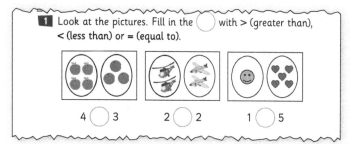
1. Look at the pictures. Fill in the ◯ with > (greater than), < (less than) or = (equal to).

4 ◯ 3   2 ◯ 2   1 ◯ 5

What learning will pupils have achieved at the conclusion of Question 1?

- Pupils will have practised completing number statements using the >, = and < symbols.

Activities for whole-class instruction

- Shuffle ten cards that show quantities similar to these:

- Choose one. Ask pupils what number it shows. Select another card and discuss whether it is greater than or less than the first card. Write the correct number statement using > or <.
- Complete Question 1 in the Practice Book, filling in the correct mathematical symbol, >, < or =.

Same-day intervention

- Place two PE mats on the floor and ask ten pupils to walk around them. Explain that when you clap your hands, they should step onto the mat that is nearest to them. Count together how many pupils are on each mat. Use large >, = and < symbols to place the correct symbol between the mats.

 The number of children on this mat is greater than the number of children on this mat.

Vary the 'All say … ' to include 'less than' and 'equal to' statements.

Same-day enrichment

- Ask pupils to explain if the age of their siblings is greater than or less than their age, for example: *Jack says, "I am 6. My brother, Toby, is 7 and my sister, Ava, is 3."*

7 > 6   3 < 6

Question 2

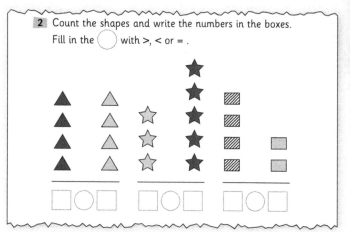
2. Count the shapes and write the numbers in the boxes. Fill in the ◯ with >, < or = .

What learning will pupils have achieved at the conclusion of Question 2?

- Pupils will have compared numbers to 10 and identified the bigger/smaller number.
- Pupils will have practised completing number statements using the >, = and < symbols.

Activities for whole-class instruction

- Place two trays of interlocking cubes at the front of the class. Invite a pupil to take a handful of cubes from one tray, make a tower and count them. Write this number on the left of the board. Ask a second pupil to do the same from the second tray and write this on the board leaving a space for the missing symbol, for example:

7   5

Discuss which is the bigger number and how the sentence will read. *7 is the bigger number. 7 is greater than 5.* Add the correct symbol to make the statement.

7 > 5

 7 is greater than 5.

Repeat with other numbers. Make sure that >, < or = are all explored.

- When you judge that pupils are sufficiently confident, give them mini whiteboards and ask them to record the number statements themselves as the cube towers are built.
- Complete Question 2 in the Practice Book, filling in the numbers and the correct mathematical symbol: >, < or =.

Chapter 1 Numbers up to 10

Unit 1.13   Practice Book 1A, pages 29–30

Same-day intervention

- Show pupils a number statement in words, for example: **Two is less than four.**
- Ask them to count out cubes and write the statement in symbols.

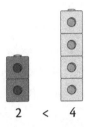

Same-day enrichment

- Use 1–10 number cards instead of interlocking cubes so pupils will have to visualise the relative size of numbers. Shuffle the cards and choose one. Select a second card and determine if it is greater than or less than the first card. Write the correct number statement using > or <.

Question 3

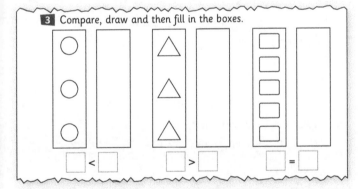

What learning will pupils have achieved at the conclusion of Question 3?

- Pupils will have compared numbers to 10 and identified the bigger/smaller number.
- Pupils will have practised completing number statements using the >, = and < symbols.

Activities for whole-class instruction

- Here are the numbers of pets that belong to pupils in Class 1.

hamster	goldfish	parrot	dog	cat
4	8	1	4	7

- Compare the following:
  - the number of hamsters with the number of each of the other pets
  - the number of goldfish with the number of each of the other pets
  - the number of dogs with the number of each of the other pets.
- Complete Question 3 in the Practice Book, filling in the numbers, mathematical symbols and shapes.

Same-day intervention

- Work backwards! Read out a number statement using >, < or =. Ask pupils to write it down on a mini whiteboard. Then find counters or cubes to show the statement.
  Repeat with another number statement.

Same-day enrichment

- In pairs, pupils make up a number statement for their partner to illustrate.

Chapter 1 Numbers up to 10

Unit 1.13 Practice Book 1A, pages 29–30

Question 4

> 4 Fill in the ◯ with >, < or =.
> 4 ◯ 6   8 ◯ 5   7 ◯ 4   9 ◯ 9
> 8 ◯ 2   7 ◯ 8   6 ◯ 6   0 ◯ 10

What learning will pupils have achieved at the conclusion of Question 4?

- Pupils will have practised completing number statements using the >, = and < symbols.
- Pupils will have compared numbers to 10 and identified the bigger/smaller number.

Activities for whole-class instruction

- Use a 1–10 spinner to generate two different numbers.
- Ask pupils to write and read number statements using the numbers, for example 4 and 8.

  4 < 8   4 is less than 8
  8 > 4   8 is greater than 4

- Complete Question 4 in the Practice Book, filling in the correct mathematical symbol, >, < or =.

Same-day intervention

- Divide pupils into three groups. Give two groups a set of 0–10 number cards. The third group will decide the correct symbol to link the numbers.

Group 1	Group 3	Group 2
3	Choose the correct symbol from >, = and < to make the statement true.	8

- Numbers can only be used once. Keep playing until all the numbers have been used.

Same-day enrichment

- Here are the numbers of big cats in a zoo.

lions	tigers	leopards	cheetahs	jaguars
10	6	3	4	8

- Ask pupils if the following statements are true or false.
  - number of leopards > number of tigers   (F)
  - number of lions > number of cheetahs   (T)
  - number of tigers < number of jaguars   (T)
  - number of cheetahs > number of tigers   (F)
  - number of jaguars < number of lions   (T)

Question 5

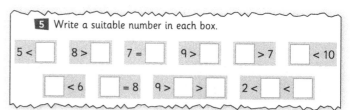

What learning will pupils have achieved at the conclusion of Question 5?

- Pupils will have practised comparing numbers mentally and using >, < and = to write number statements.
- Pupils will have practised completing number statements using the >, = and < symbols.
- Pupils will have compared numbers to 10 and identified the bigger/smaller number.

Activities for whole-class instruction

- Give pupils mini whiteboards. Ask them a series of questions such as: *Write a number smaller than 8; a number bigger than 5; a number bigger than 9; a number smaller than 3; a number bigger than 2; a number smaller than 6.*
- Share their answers and discuss the range of possible numbers for each question.
- Complete Question 5 in the Practice Book, filling in the boxes with suitable numbers. Ask pupils to explain the range of numbers possible.

Chapter 1 Numbers up to 10 Unit 1.13   Practice Book 1A, pages 29–30

Challenge and extension question

Question 6

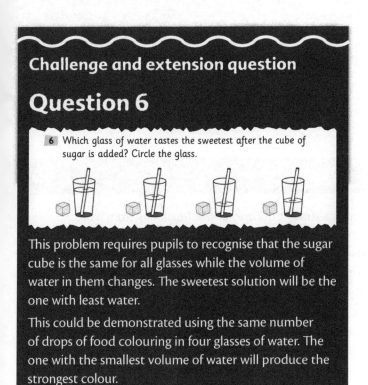

This problem requires pupils to recognise that the sugar cube is the same for all glasses while the volume of water in them changes. The sweetest solution will be the one with least water.

This could be demonstrated using the same number of drops of food colouring in four glasses of water. The one with the smallest volume of water will produce the strongest colour.

Chapter 1 Numbers up to 10

Unit 1.14
The number line

Conceptual context

In this unit, pupils are introduced to number lines. A number line is a straight horizontal line, beginning from 0 on the left, with numbers increasing in even increments along its length. The arrow at the end shows that numbers continue beyond those shown. Number lines illustrate to pupils very clearly how numbers increase incrementally. They can place their finger on the line and count the intervals to appreciate how numbers grow. Using number lines enables pupils to practise the important skill of ordering numbers.

Addition and subtraction using a number line is introduced in Chapter 2. Number lines are a valuable mathematical tool for visualising and solving problems. In due course, number lines will offer learning opportunities for higher-order skills such as negative numbers, fractions and decimals.

Counting lines

Until this point, pupils have been using counting lines, based on counting – these are also known as number tracks. These start at 1. It is not possible to insert another square at the start of a counting line and label it 0, because in counting, the first object is 1, not 0.

Number lines

In Key Stage 1, we only introduce positive numbers so they start at zero and continue to the right towards infinity. This is why they are drawn with an arrow to indicate they are not terminated at the point where the numbering ends. Older pupils will work with the number line to the left of zero.

A number line is not a ruler, so the actual distance between each number does not matter as long as each space is equal. (A number ladder is the vertical version of a number line.)

Learning pupils will have achieved at the end of the unit

- Pupils will have been introduced to number lines, developing understanding that they begin at zero, increase to the right and have no endpoint (Q1)
- Pupils will have learned that one unit on a number line is the distance between any two neighbouring numbers, for example from 0 to 1, or from 3 to 4 (Q1)
- Pupils will have practised using a number line, counting the intervals and appreciating that the intervals are evenly spaced and the numbers are in sequence (Q1, Q2, Q3, Q4)
- Pupils will have learned that you can begin counting from any position on a number line (Q3)
- Pupils will have explored making jumps of different sizes on a number line (Q3)
- Pupils will have begun to develop their ability to order numbers by visualising relationships from number symbols alone (Q4)

Resources

0–10 number line (class and individual); blank number line (class and individual); chalk; 0–10 number cards; mini whiteboards

Vocabulary

number line, unit, interval

Chapter 1 Numbers up to 10

Unit 1.14 Practice Book 1A, pages 31–33

Question 1

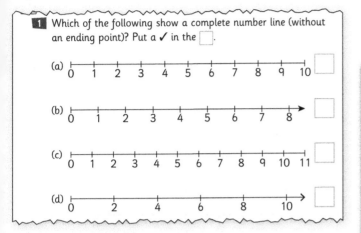

What learning will pupils have achieved at the conclusion of Question 1?

- Pupils will have been introduced to number lines, developing understanding that they begin at zero, increase to the right and have no endpoint.
- Pupils will have learned that one unit on a number line is the distance between any two neighbouring numbers, for example from 0 to 1, or from 3 to 4.
- Pupils will have practised using a number line, counting the intervals and appreciating that the intervals are evenly spaced and the numbers are in sequence.

Activities for whole-class instruction

- Look at a class 0–10 number line together. Ask pupils to point to the start and establish that the number line starts at 0. Explain that when you count with a number line, you are counting the jumps. Demonstrate how counting to 4 requires four jumps. Counting to 7 requires seven jumps and so on. Give pupils individual number lines and ask them to practise using their pointing finger to make jumps while counting the intervals.

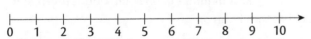

Ask pupils why the line continues with an arrow after 10. Establish that numbers do not stop at 10 and the arrow shows that numbers continue, 11, 12, 13 … A complete number line does not have an ending point.

- Show pupils a second 0–10 number line, this one marked in intervals of 2.

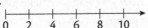

Ask them to discuss with a partner how the two lines differ. Practise counting together in twos to 10. Ask pupils to point to the place on the line where 1 would be if it was marked. What about 3? Just because they are not shown does not mean they are not there.

- Complete Question 1 in the Practice Book, ticking the complete number lines.

Same-day intervention

- Draw a chalk 0–10 number line on the playground. Remind pupils that a number line begins at 0 and has an arrow to indicate it can continue to higher numbers. Use the number line to allow them to jump along it.
- Ask pupils to spot what is wrong with these number lines.

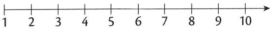

The number line does not begin at zero.

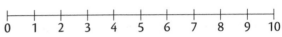

The number line has an endpoint. It should continue without an ending point.

- Point to a number on the number line and ask pupils to say where you will find a bigger/smaller number.

Same-day enrichment

- In pairs, challenge pupils to draw their own chalk 0–10 number line on the playground. Check that their number line begins at 0 and has an arrow to indicate higher numbers. Are the labels evenly spaced?

Question 2

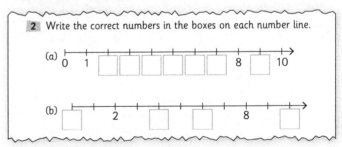

What learning will pupils have achieved at the conclusion of Question 2?

- Pupils will have practised using a number line, counting the intervals and appreciating that the labels are evenly spaced and the numbers are in sequence.

Chapter 1 Numbers up to 10

Unit 1.14 Practice Book 1A, pages 31–33

Activities for whole-class instruction

- Show pupils a blank number line and work together to complete the numbers correctly.
- Show pupils the following number lines. Spot and explain the mistakes in them.

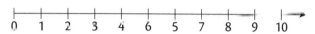

The numbers are not in the correct order. 6 and 5 are the wrong way around.

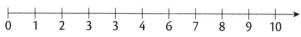

There are 2 threes in the number line.

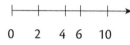

The intervals are not the same size. 4 and 6 are too close together.

- Complete Question 2 in the Practice Book, filling the boxes with the correct numbers.

Same-day intervention

- Ask pupils to look at these number lines and fill in the missing numbers.

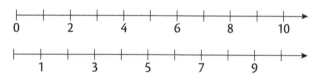

Same-day enrichment

- Draw a blank number line. In pairs, both pupils should shuffle 0–10 number cards and race to position them correctly.

Question 3

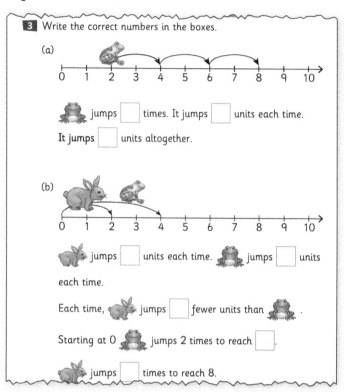

What learning will pupils have achieved at the conclusion of Question 3?

- Pupils will have practised using a number line, counting the intervals and appreciating that the intervals are evenly spaced, and the numbers are in sequence.
- Pupils will have learned that you can begin counting from any position on a number line.
- Pupils will have explored making jumps of different sizes on a number line.

Activities for whole-class instruction

- Show pupils a 0–10 number line. Ask a volunteer to start at 3 and make four jumps of one unit. Count together to see that they finish on 7.

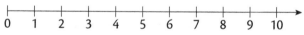

Ask where they predict they will finish if they make four jumps of one unit, starting from 4. Check. What if they start from 2? Again check to see if their prediction is correct.

- Tell pupils that now you want to make jumps of two units. Try making two jumps of two units, starting at: 0, 4, 2, 5.
- Ask pupils to describe their actions, using full sentences. For example: *I started at 4 and made 2 jumps of 2 units. I finished at 8.*

62

Chapter 1 Numbers up to 10

Unit 1.14 Practice Book 1A, pages 31–33

- Complete Question 3 in the Practice Book. Look carefully at the jumps the frog and the rabbit make and complete the blanks.

Same-day intervention

- Make the following cards.

| jumps of 1 unit | jumps of 2 units | jumps of 3 units | start at 1 | start at 0 | start at 2 |

| make 1 jump | make 2 jumps | make 3 jumps |

Shuffle the cards and place them face down. Three players choose one card of each colour and carry out the instructions. The person furthest along the number line scores a point. Repeat.

Same-day enrichment

- Pairs of pupils should take turns to throw a four-sided dice and move back along the number line from 10 to 0. If they throw a number that is too high, they miss a turn. The winner is the player who throws exactly the right number to land on 0.

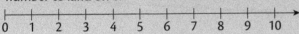

Question 4

4 Look at this number line.

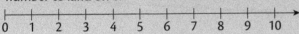

(a) Circle the numbers 7, 0, 3, 2 and 9 on the number line.

(b) Put the same five numbers in order, starting with the smallest.

☐ ☐ ☐ ☐ ☐

(c) Put the numbers 1, 8, 5, 10 and 6 in order, starting with the largest.

☐ ☐ ☐ ☐ ☐

What learning will pupils have achieved at the conclusion of Question 4?

- Pupils will have practised using a number line, counting the intervals and appreciating that the labels are evenly spaced and the numbers are in sequence.
- Pupils will have begun to develop their ability to order numbers by visualising relationships from number symbols alone.

Activities for whole-class instruction

- Challenge pupils to draw their own 0–10 number line, on a mini whiteboard, with five missing numbers. Exchange whiteboards with a partner who should complete the number line. For example:

- Use the number line to complete Question 4 in the Practice Book.

Same-day intervention

- Here are some number lines, showing jumps. Ask pupils to work out the start and finish numbers for each number line, and the size of the jump.

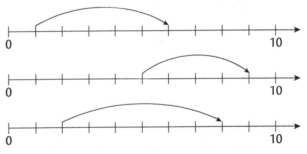

The jump starts on 2 and finishes on 8. It is 6 units.

Same-day enrichment

- Challenge pupils to order numbers without the support of a number line:
  - 6, 9, 2, 5 and 1, starting with the smallest number
  - 10, 7, 0, 4 and 8, starting with the largest number.

Check their order on the number line.

Chapter 1 Numbers up to 10

Unit 1.14 Practice Book 1A, pages 31–33

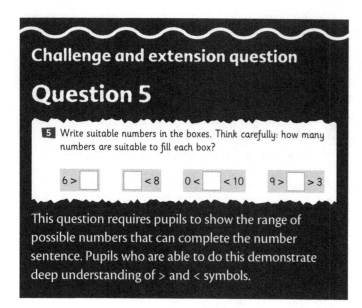

This question requires pupils to show the range of possible numbers that can complete the number sentence. Pupils who are able to do this demonstrate deep understanding of > and < symbols.

Chapter 1 test (Practice Book 1A, pages 34–38)

Test question number	Relevant unit	Relevant questions within unit
1	1.1	1
2	1.1	2, 3
	1.4	1
	1.5	1
3	1.2	2, 3, 4
	1.3	1, 2, 3
4	1.10	1, 2
	1.11	1
5	1.10	3, 4
	1.11	1, 4
6	1.11	2, 3
7	1.11	5
8	1.14	2
9	1.14	3
10	1.12	1, 2, 3, 4
11	1.13	1, 2, 3, 4, 5
12	1.13	5

Chapter 2
Addition and subtraction within 10

Chapter overview

Area of mathematics	National Curriculum statutory requirements for Key Stage 1	Shanghai Maths Project reference
Number – number and place value	Year 1 Programme of study: Pupils should be taught to:	
	■ count to and across 100, forwards and backwards, beginning with 0 or 1, or from any given number	Year 1, Units 2.1, 2.2, 2.3, 2.4, 2.5, 2.6, 2.7, 2.8, 2.9, 2.10, 2.11, 2.12, 2.13, 2.14, 2.15
	■ count, read and write numbers to 100 in numerals; count in multiples of twos, fives and tens	Year 1, Units 2.1, 2.2, 2.3, 2.4, 2.5, 2.6, 2.7, 2.8, 2.9, 2.10, 2.11, 2.12, 2.13, 2.14, 2.15
	■ identify and represent numbers using objects and pictorial representations including the number line, and use the language of: equal to, more than, less than (fewer), most, least	Year 1, Units 2.1, 2.2, 2.3, 2.4, 2.5, 2.6, 2.7, 2.8, 2.9, 2.10, 2.11, 2.12, 2.13, 2.14, 2.15
	■ read and write numbers from 1 to 20 in numerals and words.	Year 1, Units 2.1, 2.2, 2.3, 2.4, 2.5, 2.6, 2.7, 2.8, 2.9, 2.10, 2.11, 2.12, 2.13, 2.14, 2.15

Area of mathematics	National Curriculum statutory requirements for Key Stage 1	Shanghai Maths Project reference
Number – addition and subtraction	Year 1 Programme of study: Pupils should be taught to:	
	■ read, write and interpret mathematical statements involving addition (+), subtraction (−) and equals (=) signs	Year 1, Units 2.1, 2.2, 2.3, 2.4, 2.5, 2.6, 2.7, 2.8, 2.9, 2.10, 2.11, 2.12, 2.13, 2.14, 2.15
	■ represent and use number bonds and related subtraction facts within 20	Year 1, Units 2.1, 2.2, 2.3, 2.4, 2.5, 2.6, 2.7, 2.8, 2.9, 2.10, 2.11, 2.12, 2.13, 2.14, 2.15
	■ add and subtract one-digit and two-digit numbers to 20, including zero	Year 1, Units 2.1, 2.2, 2.3, 2.4, 2.5, 2.6, 2.7, 2.8, 2.9, 2.10, 2.11, 2.12, 2.13, 2.14, 2.15
	■ solve one-step problems that involve addition and subtraction, using concrete objects and pictorial representations, and missing number problems such as 7 = ☐ − 9.	Year 1, Units 2.4, 2.5, 2.6, 2.7, 2.8, 2.9, 2.13, 2.14
Number – number and place value	Year 2 Programme of study: Pupils should be taught to:	
	■ compare and order numbers from 0 up to 100; use <, > and = signs.	Year 2, Units 2.12, 2.15
Number – addition and subtraction	Year 2 Programme of study: Pupils should be taught to:	
	■ add and subtract numbers using concrete objects, pictorial representations, and mentally, including: adding three one-digit numbers	Year 2, Units 2.13, 2.15
	■ show that addition of two numbers can be done in any order (commutative) and subtraction of one number from another cannot	Year 2, Units 2.2, 2.3, 2.4, 2.5, 2.6, 2.10, 2.12, 2.13
	■ recognise and use the inverse relationship between addition and subtraction and use this to check calculations and solve missing number problems.	Year 2, Units 2.7, 2.10, 2.11, 2.12

Chapter 2 Addition and subtraction within 10

Unit 2.1
Number bonds

Conceptual context

This is the first in a series of units on addition and subtraction within 10. The focus is on recognising that each number can be split (partitioned) into parts, with a focus on two parts because these form the basis of number bonds. Other partitioning will be explored in subsequent chapters.

At this stage, the focus is on beginning with the whole and splitting it into two parts. This can be done in several different ways. Pupils are shown how to work systematically so that they can be confident they have found all the possibilities. As pupils become more familiar with this approach, they will explore how to use the set of objects to identify a missing quantity. The quantity of objects is small enough to give pupils ample opportunity to practise subitising. It is important that pupils develop the skill of working systematically since this will be of use throughout mathematics. It will support their growing knowledge by exposing patterns which they can then internalise and apply in other situations.

The language used to verbalise the part–whole relationship is developed in stages into the language of addition. Once this is introduced, the symbols for writing an addition number sentence are also introduced. Pupils are not yet calculating since they are manipulating a physical quantity or drawings to identify an unknown quantity. Towards the end of this unit, pupils will be beginning to calculate if they can complete a partitioning tree or number sentence without the need to model it first.

It is important that, through these activities and questions, pupils have the opportunities to learn that:

(a) a quantity can be partitioned into smaller amounts; in other words, they discover that smaller numbers are included 'within' the larger whole

(b) working systematically ensures that they can find all the possible solutions and be confident that they have found them all.

Learning pupils will have achieved at the end of the unit

- Pupils will have been introduced to the underlying patterns of partitioning numbers to 10 (Q1)
- Pupils will have identified all the possible combinations of parts of a number by working systematically, and will be able to justify how they know they have found them all (Q1)
- Pupils will have used subitising to identify parts and wholes (Q1)
- Pupils will have further developed their understanding of part–whole relationships (Q2)
- Pupils will have consolidated their understanding of the use of abstract tokens to represent objects (Q2)
- Pupils will have explored recording part–whole relationships in abstract formats such as partitioning trees (Q2)
- Pupils will have begun to develop strategies to identify the missing number or numbers in a partitioning tree (Q2)
- Pupils will have consolidated their recording of part–whole relationships as number bonds in an abstract format (Q3)
- Pupils will have explored how to complete the unknown part or parts of a partitioning tree by relating it to a part–whole statement and number bond (Q3)

Resources

two PE mats; counting objects including counters, cubes, buttons, pebbles, conkers and so on; mini whiteboards; paper plates; tablets or cameras ; 0–10 number cards

Vocabulary

0, 1, 2, 3, 4, 5, 6, 7, 8, 9, 10, part, whole, and, is, altogether, add, equals, partition, partitioning tree, sum

Chapter 2 Addition and subtraction within 10

Unit 2.1 Practice Book 1A, pages 39–40

Question 1

> **1** Complete the tables. One has been done for you.
>
> | ○○○○ | 0 | 4 |
> |---|---|---|
> | ●○○○ | | |
> | ●●○○ | | |
> | ●●●○ | | |
> | ●●●● | | |
>
> | ●○○○○ | | |
> |---|---|---|
> | ●●○○○ | | |
> | ●●●○○ | | |
> | ●●●●○ | | |
> | ●●●●● | | |

What learning will pupils have achieved at the conclusion of Question 1?

- Pupils will have been introduced to the underlying patterns of partitioning numbers to 10.
- Pupils will have identified all the possible combinations of parts of a number by working systematically, and will be able to justify how they know they have found them all.
- Pupils will have used subitising to identify parts and wholes.

Activities for whole-class instruction

- Set out two PE mats with four pupils on one mat and none on the second mat. Ask pupils to say what they see in a sentence, for example: *There are four children on one mat and none on the other.* Model back to pupils: *The whole is 4. One part is 4, the other part is 0.* All repeat together.
- Move one child to the empty mat.

   All say… The whole is 4, one part is 3, the other part is 1.

  Continue moving one pupil at a time and verbalising each new arrangement until all four pupils are on the previously empty mat.
- Discuss what would happen if there were a different number of pupils on the starting mat. Repeat with five pupils if further reinforcement is needed.
- Ask pupils to work in pairs and get 6, 7, 8, 9 or 10 counting objects and two paper plates. Provide pupils with a tablet or camera to photograph each step. Pupils treat the paper plates as mats and explore moving objects from one plate to the other to find all the different ways of partitioning their chosen number. Remind pupils that they may be able to subitise small quantities rather than have to count objects individually.

  Look out for … pupils who work systematically, beginning with all their counters on one plate and moving them one at a time to the second plate. Ask a pair of pupils who worked systematically to display their photos on the whiteboard.

- Give pupils the opportunity to order their photos if they did not work systematically.
- Give pupils interlocking cubes in two colours to model the systematic pattern for their chosen number. Pupils display their patterns on a mini whiteboard, recording the parts alongside. Photograph one version for each of the numbers from 6 to 10 for display.
- Use one or two of the display photographs to support verbalising the number bonds for the number, using your chosen format of parts and whole, for example 1 is a part, 5 is a part, 6 is the whole. Extend to 1 and 5 is 6 altogether.
- When pupils explore the tables in the Practice Book, ask them to tell you what is missing from the second table. Compare the two tables and identify that the second table does not have a row of all white counters. Ask pupils to draw in the missing row.

(i) Knowledge of number bonds to 10 is a vital mental calculation tool. Pupils need repeated concrete experience of these to begin to embed them so that they can remember and apply them. Pupils have begun to get to know each number to 10 as a whole quantity. Each number bond is that particular quantity partitioned into two groups or parts. Every number has its own unique and intrinsic set of number bonds, but every set can be explored systematically and similar patterns noted. The patterns can then be used to support recall.

Same-day intervention

- Give pupils a copy of an empty grid with 11 rows, sufficient to work with numbers up to 10 in the same style as that in Question 1 in the Practice Book, and counters in two different colours or double-sided counters. Agree a number between 6 and 10. Pupils begin by placing that many counters all the same colour on the first row. Label the top of the second and third columns with the relevant colour. Pupils record how many of that colour and how many of the second colour in these columns. Encourage pupils to subitise. Work systematically in each subsequent row, changing one counter at a time to the second colour and recording how many of each colour. Explore how one part increases as the other part decreases. Compare with a different number, exploring what is the same and what is different.

Chapter 2 Addition and subtraction within 10

Unit 2.1 Practice Book 1A, pages 39–40

Same-day enrichment

- Ask pupils to explain how they know they have found all the possible combinations of parts. Pupils may successfully complete the activity but find it difficult to explain how they know they have all the possible numbers of parts.

Question 2

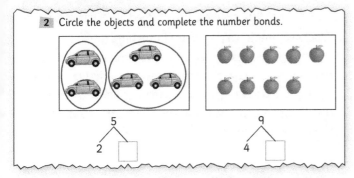

What learning will pupils have achieved at the conclusion of Question 2?

- Pupils will have further developed their understanding of part–whole relationships.
- Pupils will have consolidated their understanding of the use of abstract tokens to represent objects.
- Pupils will have explored recording part–whole relationships in abstract formats such as partitioning trees.
- Pupils will have begun to develop strategies to identify the missing number or numbers in a partitioning tree.

Activities for whole-class instruction

- Show pupils a set of six objects on the whiteboard, set out in two parts, 2 and 4. Ask: *What is the whole? What is one of the parts? What is the other part?*
- Read the displayed image together, for example 6 is the whole, 2 is a part, 4 is a part. Add a blank partitioning tree alongside the image. Place six objects in the top box of the tree and ask pupils how to make the partitioning tree show the same as the original image. Complete the image and then call up a second partitioning tree. Complete this together using numbers instead of objects. Ask pupils how each image is the same and how they are different. Repeat the description: *6 is the whole, 2 is a part, 4 is a part*, to reinforce the fact that this is a further image of the same thing.

- Give pupils mini whiteboards and ask them to draw seven simple images. Display a partitioning tree with '7' at the top of the tree and then write '2' in one of the parts. Ask pupils to annotate their drawing to match the tree. Count or subitise to confirm that the other part is 5. Repeat with a few more examples, varying whether you give the whole and one part or the two parts. Move on to only giving the whole.
- **Look out for** … pupils who can work systematically to list all the possible ways of completing this type of partitioning tree.
- In a group, pupils choose a number up to 10 to explore. Each pair draws and completes a partitioning tree on a mini whiteboard. Pupils then order their boards to check if they have shown all the possible parts and complete further boards for any missing parts.
- Ask one group to show their set of boards to the rest of the class. Pupils read each board using the part–whole format, with the group checking that they are correct. Model the number statement format: *2 and 5 equals 7* and then all read the set again in that format. Explain 'equals' means 'is the same as' or 'has the same value as'.

Same-day intervention

- Draw a large partitioning tree on paper or a mini whiteboard. Choose a number such as 6 to explore. Ask a pupil to place six objects in the top of the tree. Move all the objects into one of the part spaces, saying: *If we put 6 here and 0 here, how many have we got altogether?* Return the 6 to the top of the tree and move 1 into one of the part spaces. The pupil moves the rest into the other part space and says what they see. Encourage pupils to subitise. Return the objects to the top of the tree and move 2 into one of the part spaces. Continue in the same way until all the part–whole statements for 6 have been modelled practically and verbally. If necessary, repeat for another quantity.

Same-day enrichment

- Ask pupils to produce a set of partitioning trees in their chosen format for each of the numbers 6 to 10. Challenge pupils to explore how many trees there are for each number and to explain what they notice. The number of trees is always 1 more than the chosen number, because parts range from 0 to the number being considered, 1 more than the number itself. Ask them if they find it easier when they move one object at a time from one space to another – why?

Chapter 2 Addition and subtraction within 10

Unit 2.1 Practice Book 1A, pages 39–40

Question 3

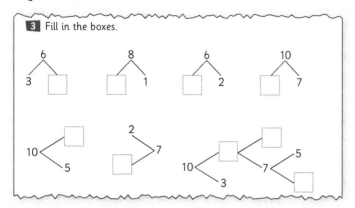

What learning will pupils have achieved at the conclusion of Question 3?

- Pupils will have consolidated their recording of part–whole relationships as number bonds in an abstract format.
- Pupils will have needed to think flexibly to find missing elements of partitioning trees.

Activities for whole-class instruction

- Give pairs of pupils a set of number cards, 4–10. Pupils shuffle the cards and turn over the top card. They draw a partitioning tree on a whiteboard, placing the number card in the whole position of the tree. They then complete the partitioning tree using concrete resources, drawings or numerals. Pupils take turns to read their partitioning tree to their partner using and, is or equals.
- When pupils explore Question 3 in the Practice Book, some may get stuck when they come to the branching tree.

 ... pupils who found this straightforward.

- Display the tree one branch at a time, working through it together to identify missing numbers with the support of those pupils who found it straightforward. Breaking the question down and focusing on one branch without the distraction of further branches will support those pupils who found this difficult.
- Introduce + (add) and = (equals) as symbols for recording without drawing boxes. Return to some of the questions in the Practice Book and verbalise them using the part–whole and the and/is or equals format, recording some in horizontal format with + and =. Work through the branching tree together, verbalising and recording each number sentence.

Same-day intervention

- Give pupils concrete apparatus to help them complete a partitioning tree. Extend the tree by making one of the parts the new whole and drawing two further parts. With the new whole modelled using apparatus, pupils complete the additional branches. Repeat with further examples.

Same-day enrichment

- Challenge pupils to produce a branching tree for another pupil to complete. Provide mini whiteboards so that pupils can complete the tree as they design it, then erasing the parts they want a partner to find. Extend to designing on paper, without the benefit of completing the tree as it is designed.

Challenge and extension question

Question 4

4 Think carefully and then fill in the boxes with suitable numbers.

(a) $5 + 1 = 4 + 2 = \square + \square = \square + \square$
      $= \square + \square$

(b) $9 + 1 = 8 + 2 = \square + \square = \square + \square$
      $= \square + \square = \square + \square = \square + \square$
      $= \square + \square = \square + \square = \square + \square$

This question requires pupils to recognise that the number sentences they have been verbalising can be recorded in a more abstract format, using symbols for the words add and equals as well as numerals for numbers. Some pupils will find it helpful to identify the whole and then draw the set of partitioning trees for that whole before completing the recording. If they use this approach, they may notice that there are not enough spaces to record all the possibilities for 6. Pupils could extend the template so that they record them all. Challenge pupils to record all the possible additions for a different number in the same format.

Chapter 2 Addition and subtraction within 10

Unit 2.2
Addition (1)

Conceptual context

This is the second in a series of units on addition and subtraction within 10. Having been introduced to the idea that numbers can be partitioned into two parts in different ways, this unit focuses on identifying and recording the parts and the whole as addition sentences.

After revising a whole split into two parts using a partitioning tree, the unit moves on to identifying the parts from an illustration. Pupils record the parts within a number sentence and find the sum. At this point, pupils could subitise or use the illustration to count all or count on from one of the parts to find the sum. The key idea that the parts can be recorded in any order is introduced through recording two number sentences for each illustration. The language of addition is developed with the introduction of 'addend' for the parts of the whole and the reinforcing of 'sum' for the whole. The commutative law is introduced in this unit; pupils learn that changing the order of addends does not change the sum.

In later questions, quantities are represented only as abstract numerals, rather than as illustrations of a countable number of objects, so pupils are encouraged to calculate rather than count.

The unit moves on to providing pairs of addends for pupils to quickly calculate and match. This refines the idea that if a set of addends have the same sum, they are linked through that sum. Understanding is both developed and deepened by showing number sentences as a balance on a set of balance scales. When the scales do not balance, the number sentence is incorrect.

It is important that, through these activities and questions, pupils have the opportunities to learn that:
- part–whole relationships can be recorded in a variety of different formats
- addends (parts) can be recorded in any order; the sum (whole) remains the same.

Learning pupils will have achieved at the end of the unit

- Pupils will have consolidated recording part–whole relationships in abstract formats such as partitioning trees (Q1)
- Pupils will have recognised that partitioning trees can be drawn in different orientations and styles but still give the same information (Q1)
- Pupils will be able to use subitising to quickly identify the quantity value of parts and wholes in an illustration (Q2)
- The commutative nature of addition will have been encountered and developed through concrete experience and pictorial representations (Q2, Q3)
- Pupils will have explored recording illustrated part–whole relationships in an abstract format such as a number bond (Q2)
- Pupils will have reinforced the link between parts and wholes through number bonds (Q2)
- Pupils will have been introduced to 'addend' for the parts in a part–whole relationship and reinforced the use of 'sum' for the whole (Q3)
- Pupils will have explored recording part–whole relationships in an abstract format within a table, reinforcing the link between parts and wholes through recording in columns (Q3)
- Understanding that different pairs of numbers can be added together to achieve the same sum will have been reinforced (Q4)
- Pupils will have explored equivalence of number bonds through modelling (Q4)

Resources

mini whiteboards; counting objects such as counters, cubes and small toys; adult and baby animal models and pictures; 0–10 number cards; balance scales; interlocking cubes; toy trucks and cars; timer

Vocabulary

whole, part, relationship, partitioning, partitioning tree, top, bottom, left, right, number sentence, number bond, add, +, equals, =, addend, sum, balance, value, subitise

Chapter 2 Addition and subtraction within 10 Unit 2.2 Practice Book 1A, pages 41–42

Question 1

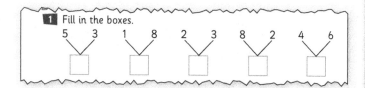

What learning will pupils have achieved at the conclusion of Question 1?

- Pupils will have consolidated recording part–whole relationships in abstract formats such as partitioning trees.
- Pupils will have recognised that partitioning trees can be drawn in different orientations but still give the same information.

Activities for whole-class instruction

- Draw the partitioning tree as shown. Ask pupils to tell you what you have drawn. Confirm that it is a partitioning tree to show that 8 is the whole, 5 is a part and 3 is a part. Give pupils a range of counting objects to model the partitioning tree.
- Ask pupils to work in pairs to draw some more partitioning trees for the same part–whole relationship, 5 is a part, 3 is a part and 8 is the whole. Share ideas, displaying examples with the whole at the top, bottom and left or right. Pupils may also place the whole and parts at different angles to each other. Accept all arrangements/orientations provided it is clear which is the whole and which are the parts.
- Highlight the four standard arrangements with the whole at the top, bottom, left or right. Confirm that these are the versions usually seen in books.

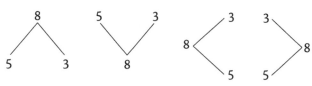

- Complete Question 1 in the Practice Book.

Same-day intervention

- Draw a partitioning tree with 8 at the top and the two parts 5 and 3 below on a whiteboard. Ask pupils to take eight counters and partition the 8 into groups of 5 and 3. Read the partitioning together, using the tree and counters for support: *8 is the whole, 5 is a part and 3 is another part.*
- Give each pupil a mini whiteboard. Rotate the displayed partitioning tree a quarter turn. Ask pupils to draw the rotated partitioning tree, so that the whole is on the right, writing the numbers upright. Ask pupils to check if they need to change any part of their arrangement of counters. Read the partitioning tree together.

 8 is the whole, 5 is a part and 3 is a part.

- Check that pupils recognise that they said the same thing again – although the tree looks a little different, nothing has changed so it contains the same information as before and can be modelled with counters in exactly the same way. Rotate the displayed partitioning tree back to its original position and ask pupils to rotate their boards to match, to confirm that they are the same as the original.
- Repeat for a half turn and a three-quarter turn.

Same-day enrichment

- Give pupils a partitioning tree with the whole at the bottom and the two branches extending upwards. Challenge them to extend the tree by partitioning each part into two further parts.
- Share the extended trees that pupils have drawn. Confirm that this layout looks more like a tree than any other layout. Discuss which numbers are displayed at the ends of the tree. If pupils have persevered with their partitioning, all final branches will be 0 or 1. Challenge those who did not persist to extend their partitioning tree until the further boxes all contain 1 or 0. Ask pupils to explain why they cannot make all the final branches 0.

Chapter 2 Addition and subtraction within 10

Unit 2.2 Practice Book 1A, pages 41–42

Question 2

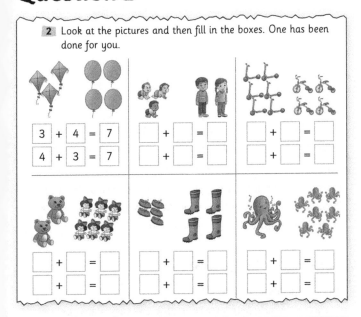

What learning will pupils have achieved at the conclusion of Question 2?

- Pupils may have used subitising to quickly identify the quantity value of parts and wholes in an illustration.
- Pupils will have explored recording illustrated part–whole relationships in an abstract format such as a number bond.
- The commutative nature of addition will have been encountered through concrete experience.
- Pupils will have reinforced the link between parts and wholes through number bonds.

Activities for whole-class instruction

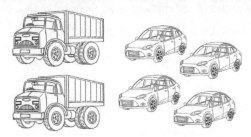

- Show pupils two trucks and four cars, encouraging pupils to subitise the quantities. Agree that the whole is vehicles, one part is trucks, the other part is cars. Read the arrangement together, for example: *6 is the whole, 2 is a part, 4 is a part.*
- Ask pupils to draw a partitioning tree for the vehicles. Confirm that trees can be drawn with the whole at the top or the bottom or to one side; the information within the tree is the same. Remind pupils that in the previous unit they started to write a number sentence to record all the ways of partitioning a number into two parts. Look at the vehicles again and record how many of each type of vehicle. Remind pupils that the 2 and the 4 are parts eight come together to make the whole, so the number sentence to say this is 2 + 4 = 6.
- Swap the two quantities over and ask pupils to say the new number sentence, 4 + 2 = 6. Record this with the first number sentence and ask pupils to talk to their partner about how this is the same and how it is different from the first number sentence. Remember to swap both representations (concrete and abstract).
- Share responses. Confirm that both number sentences describe the vehicles. Although the parts have been written in a different order, the value of each part is the same and the sum is the same: 4 and 2 are still bound together to equal 6.
- Display some similar arrangements of objects or illustrations and ask pupils to write paired number sentences for each of them, using the model on the whiteboard for support.
- Complete Question 2 in the Practice Book.

Same-day intervention

- Carry out the same activity using model animals. Pupils identify the adults and babies supplied and choose one set to put on their whiteboards, separating the adults and babies into two clear groups. Pupils count and label each group, then record the full number sentence. Verbalise the part–whole relationship in the same order as the number sentence, for example: *2 is a part, 3 is a part, 5 is the whole*: 2 + 3 = 5. Exchange the groups of animals for matching pictures and ask pupils if it changes the number sentence. Confirm that none of the quantities have changed; there are the same number of animals in each part and in the whole.
- Swap over the groups or pictures for pupils to record the second number sentence for their part–whole relationship. Repeat with a different set of animals and pictures.

Same-day enrichment

- Give pupils a single picture and a partly completed number sentence. For example, a picture of four puppies and the number sentence 4 + ☐ = 7. Ask pupils to complete the drawing and the number sentence.

Chapter 2 Addition and subtraction within 10

Unit 2.2 Practice Book 1A, pages 41–42

Question 3

3 Complete the table.

addend	8	2	5	1	3
addend	2	4	3	9	3
sum					

addend	7	5	1	2	6
addend	1	5	3	1	4
sum					

What learning will pupils have achieved at the conclusion of Question 3?

- Pupils will have been introduced to the word 'addend' for the parts in a part–whole relationship and reinforced the use of 'sum' for the whole.
- Knowledge of the commutative law will have been developed.
- Pupils will have explored recording part–whole relationships in an abstract format within a table, reinforcing the link between parts and wholes through recording in columns.

Activities for whole-class instruction

- Give pupils two sets of 0–5 digit cards; also give them a + and a = card. Pupils work in pairs, shuffling the digit cards and turning over the top two. They use the cards to create two addition number sentences, arranging an addition calculation and then swapping their numbers to create a second calculation. Set a timer for two minutes and challenge pupils to generate and solve as many addition sentences as they can within that time.
- When the time is up, ask pupils to check that they have recorded both versions of the same number sentences and to cross out any duplicates.
- Display a sample pair of number sentences. Explain that each part of the number sentence has its own name. Check that pupils can explain what the + and = symbols are called. Then explain that the numbers being added are called 'addends' and the total is called the 'sum'. Write: 'addend + addend = sum' below the number sentences. Confirm that both numbers to be added are called 'addend', so their order in the number sentence does not matter: the sum will be the same. Explain that the written words look like a sentence, though without the capital letter and full stop, so when we say the same thing in numbers, it is called a number sentence.
- Tell pupils that because they recorded their number sentences quickly, they are not as neat as they could be. Show them the table:

Addend					
Addend					
Sum					

Record the sample number sentences in the table. Explain that tables are a good way to record lots of information.

- Give pupils one or more copies of the table and ask them to complete it with their number sentences. Display the tables around a large copy of the sentence addend + addend = sum to remind pupils of the correct language.

Same-day intervention

- Give pupils a copy of the sentence addend + addend = sum. Shuffle the digit cards as before. Place the top two cards below the words addend. Give pupils counters or other counting objects to help them complete the number sentences. Swap the numbers over and ask pupils to work out the sum again. Confirm that the sum is the same as before.
- Support pupils to add their number sentences to a table like the one used earlier in the session. Collect the number sentences from everyone in the group in one table and add to the display.

Same-day enrichment

- Ask pupils to choose a number from 6 to 10. Give them a copy of the table and tell them that they have chosen their sum. Challenge them to complete the table with all the possible ways of adding two numbers to make that sum.
- Compare tables. Ask: *Does the greatest sum need the most columns in the table?*

Question 4

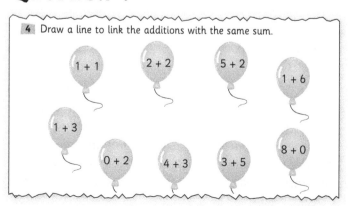

4 Draw a line to link the additions with the same sum.

Chapter 2 Addition and subtraction within 10

Unit 2.2 Practice Book 1A, pages 41–42

What learning will pupils have achieved at the conclusion of Question 4?

- Understanding that different pairs of numbers can be added together to achieve the same sum will have been reinforced.
- Pupils will have explored equivalence of number bonds through modelling.

Activities for whole-class instruction

- You will need a set of balance scales and interlocking cubes in different colours. Ask pupils to help you make towers of three red cubes, four yellow cubes and seven blue cubes.
- Write the number bond 3 + 4 = 7 on the whiteboard and ask pupils to tell you what the equals sign means. Agree that either side of the equals sign has the same value, so 3 + 4 has the same value as 7. Place the towers of three and four cubes on one side of the scales and the tower of seven on the other. Attach an equals card to the central part of the balance when it is level, explaining that we can think of the equals sign as a balance. The number sentence is true if it balances, that is if both sides of the equals sign have the same value. Add another cube to the seven blue cubes and ask if the scales still balance. Confirm that they do not, so the equals sign must be removed. 3 + 4 does not equal 8.
- Ask pupils to make a stick of seven cubes using two different colours of cubes. Remove the eight cubes from the balance scales and test some of pupils' towers against 3 + 4 to see if they balance. For those that do, write the matching number sentence, for example 3 + 4 = 5 + 2; 3 + 4 = 6 + 1. Ask pupils to work in pairs to make four towers of their own to test on the balance scales. Invite pairs to test their towers, saying the number sentence they have made.
- Complete Question 4 in the Practice Book.

Same-day intervention

- You will need cubes in two different colours. Ask pupils to make a tower of five cubes, all the same colour. Place the tower on one side of the balance. Invite pupils to add some cubes to the other side until they balance. Count the cubes together to confirm there are 5, then sort them into the two colours. Write the matching number sentence.
- Repeat with other small numbers.

Same-day enrichment

- Ask pupils to place a large handful of cubes on each pan of the balance scales. If the scales do not balance, they should move a single cube from one pan to the other until they do balance. You might need to discard one cube. (Pupils might be able to tell you why.)
- Pupils then sort the cubes in each pan into colour groups and record the matching number bond for the balanced sum.

Challenge and extension question

Question 5

5 Write addition calculations with a total of 5.

☐ + ☐ = 5   ☐ + ☐ = 5   ☐ + ☐ = 5

☐ + ☐ = 5   ☐ + ☐ = 5   ☐ + ☐ = 5

This question requires pupils to record all the possible number sentences with a total of 5 in an abstract format. There are no illustrations for support, so pupils will need to recognise the format of the number sentences and supply the addends to complete the empty boxes. Some pupils will work mentally, while others may need cubes or counters for support.

Look out for … pupils working systematically, increasing and decreasing pairs of addends by one repeatedly until all the possible solutions have been found.

Pupils will need to use all the outline sentences provided to record all the solutions. Encourage pupils to check their number sentences if they have an outline sentence left over or need to draw another one. Early finishers should record all the possible additions for a different sum in the same format. Can they predict how many solutions there will be?

Chapter 2 Addition and subtraction within 10

Unit 2.3
Addition (2)

Conceptual context

This is the third in a series of units on addition and subtraction within 10. Having been introduced to the idea that numbers can be partitioned into two parts, this unit also focuses on identifying and recording the parts and the whole in a number sentence, but the amount of support provided is gradually reduced.

After identifying and recording the parts and whole from a series of illustrations organised systematically for 8, pupils move on to partitioning the whole themselves using a similar illustration. With the demonstration of working systematically in part (a), pupils should be able to work systematically in part (b). The unit then moves on to providing clearly structured parts for pupils to subitise or count to construct the number sentence. They can then find the sum by counting all or counting on. Pupils may be beginning to recall some repeatedly encountered number bonds. Question 3 gives bare number sentences for pupils to find the sum. Since there is nothing to count, pupils should count on from the first addend. Some pupils may recognise that addends can be added in any order, and reason that it is more efficient to count on from the higher number. The challenge and extension question returns to providing an illustration, but allows pupils to choose their own combinations of addends.

The focus of the unit is on continuing to reduce the support provided for identifying and recording a number sentence. This includes presenting addends without illustrations, which encourages pupils to calculate rather than count.

It is important that, through these activities and questions, pupils have the opportunities to learn that:
- part–whole relationships can be recorded in number sentence format
- by working systematically, all the possible number sentences for one sum can be identified.

Learning pupils will have achieved at the end of the unit

- Pupils will have consolidated recording part–whole relationships using symbols in a number sentence with and without structural support (Q1, Q2)
- Pupils will have consolidated conceptual connections between parts and wholes through number sentences (Q1, Q2, Q3)
- Pupils will have learned how to work systematically to find all possibilities and be able to explain how they know that they have found all possible solutions (Q1)
- Pupils will have again been encouraged to subitise to quickly identify the quantity value of parts and wholes in an illustration (Q2)
- Pupils will have recorded part–whole relationships presented as images as number sentences with and without structural support (Q2)

Resources

double-sided counters; dice; counters; addition pyramids

Vocabulary

addend, sum, number bond, systematic, addition sentence

Chapter 2 Addition and subtraction within 10

Unit 2.3   Practice Book 1A, pages 43–45

Question 1

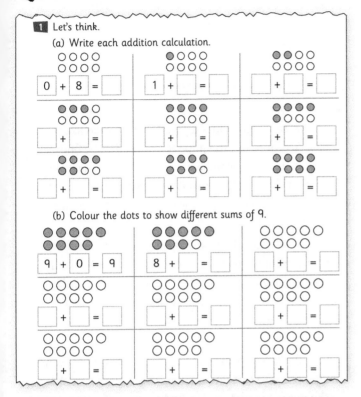

What learning will pupils have achieved at the conclusion of Question 1?

- Pupils will have consolidated recording part–whole relationships using symbols in a number sentence.
- Pupils will have consolidated conceptual connections between parts and wholes through number sentences.
- Pupils will have worked systematically to find all possibilities and will be able to explain how they know that they have found all possible solutions.

Activities for whole-class instruction

- Give each pair of pupils five double-sided counters and ask them to arrange the counters in a single row, all showing the same colour, for example red.
- Ask pupils: *If 5 is the whole, what are the parts?* Say the word and number sentences together, for example: *0 is a part, 5 is a part and 5 is the whole, 0 + 5 = 5.* Record the number sentence.
- Ask pupils to turn over the first counter to show its second colour, for example blue, and describe what they see to their partner.

   1 is a part, 4 is a part and 5 is the whole, 1 + 4 = 5.

- Record the number sentence below the counters, lining up the numbers with two groups of counters. Ask pupils what the 1 represents (the blue counter) and what 4 represents (the red counters) and what the 5 represents (the sum or whole). Also begin to make a list of the number sentences.
- Continue in the same way, turning over each counter in turn until all have changed colour. Say the word and number sentences each time, recording below the counters to reinforce the link between the concrete and the abstract. Build a list of number bonds/sentences. Occasionally, pause to check that pupils are clear about what each number represents.
- Ask pupils if they have found all the possible two-part number bonds with a sum of 5 and how they know.

- Give each pair of pupils a further counter and ask them to arrange their counters in two rows of three with all counters showing the same colour. Work systematically as before, turning over each counter on the top row then those on the bottom row until all have changed colour. Say and record the number sentences, building a list as before.
- Compare both lists, asking how they are the same and how they are different.
- When pupils complete Question 1 in the Practice Book, ask them to work systematically so that they know they have found all the possible number sentences for 8 and 9.

Same-day intervention

- Revisit the whole-class activity, but, instead of repeatedly turning over another counter, create further arrangements of six counters and turn over one then two then three counters and so on. Say the word and number sentence each time, recording the number sentence alongside each arrangement.
- When all seven arrangements of six counters have been produced, reread each number sentence including what each number represents, for example zero red counters and six blue counters equals six counters altogether; one red counter and five blue counters equals six counters altogether and so on. Show that the number bonds have used 0, 1, 2, 3, 4, 5 and 6 for each addend. Since there are no more numbers within 6 to use, all possible number bonds have been found. How many were there?

Chapter 2 Addition and subtraction within 10

Unit 2.3   Practice Book 1A, pages 43–45

Same-day enrichment

- Ask pupils to list all the possible number bonds with two parts and a sum of 6. Challenge them to match number sentences that use the same addends, for example 2 + 4 = 6 and 4 + 2 = 6. How many pairs can they find? Are there any number sentences without a pair? Why? Repeat with 7.

the arrangement that allows you to quickly identify or work out how many. For example, 7 could be shown as 4 and 3 dice patterns next to each other.

- Complete Question 2 in the Practice Book.

Same-day intervention

- Place seven counters on the table in a random arrangement. Introduce a puppet who would like to learn to count. Let the puppet explain that he finds it hard to know when he has counted something and when he hasn't. Demonstrate, counting some counters more than once and getting in a muddle so that the count trails off.
- Ask pupils how they can help the puppet. Encourage them to show him how to put objects in a line and count along touching each object as he moves along the row. Introduce a dice and ask pupils to explain how they use the spot patterns to know how many really quickly.
- Ask pupils to test the puppet by organising the seven counters into two dice arrangements. Make sure that the puppet makes some mistakes so pupils can correct him, pointing to the arrangements and reading them correctly, with your support if necessary.

Question 2

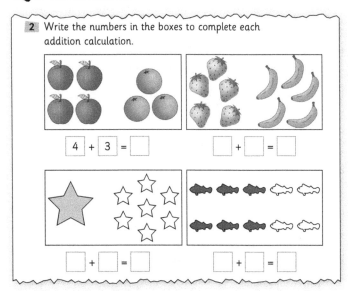

What learning will pupils have achieved at the conclusion of Question 2?

- Pupils will have practised subitising.
- Pupils will have recorded part–whole relationships presented as images as number sentences with and without structural support.
- Pupils will have continued to reinforce the link between parts and wholes through number sentences.

Same-day enrichment

- Ask pupils to make some cards with at least four different spot arrangements for 7, 8, 9 and 10. Shuffle the cards and play snap, with pupils calling out 'snap' and the number matched.
- Shuffle the cards and lay them out in a grid. Pupils take turns to turn over two cards. If they match, the pupil keeps the cards. If not, they are turned face down again. The winner is the player who collects the most cards.

Activities for whole-class instruction

- Talk about how organising counters into rows helped pupils to work systematically in the previous activity. It also helped pupils to quickly subitise quantities.
- Give pairs of pupils a dice and counters and ask them to recreate each dice pattern from 1 to 6 using counters. Pupils take turns to re-order the patterns and challenge their partner to say each quantity without hesitation. They then take turns to point to two dice patterns for their partner to say as a number bond, including the sum.
- Challenge pupils to create their own arrangements for 7, 8, 9 and 10. Share some ideas, discussing what it is about

Question 3

3 Work out the calculations.

3 + 5 =   4 + 6 =   2 + 5 =

0 + 4 =   4 + 4 =   5 + 1 =

2 + 2 =   6 + 2 =   2 + 8 =

3 + 3 =   6 + 3 =   5 + 4 =

Chapter 2 Addition and subtraction within 10

Unit 2.3   Practice Book 1A, pages 43–45

What learning will pupils have achieved at the conclusion of Question 3?

- Pupils will have practised adding using number sentences.
- Pupils will have been encouraged to recognise efficient strategies for adding by counting on from the highest addend.
- Knowledge of the commutative law for addition will have been reinforced.
- Pupils will have had further opportunities to reinforce their knowledge of number bonds to 10.

Activities for whole-class instruction

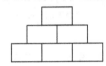

- Show pupils the pyramid with a base of three bricks. Write the numbers 1, 2 and 3 in the bottom row. Explain that the numbers in the two bricks supporting the brick above must be added together to find which number to write in the brick. Complete the pyramid together, asking pupils to explain how they know which number belongs in each brick.
- Give pupils a sheet with six pyramids on and ask them to explore rearranging 1, 2 and 3 to see if they make the top brick a different number.
- Share results, looking for different top brick numbers and the same number made in different ways. Discuss whether putting 3, 2 and 1 in the base bricks is the same as putting 1, 2 and 3.

Same-day intervention

- In addition to writing the numbers 1, 2 and 3 in the base of the pyramid, place one, two and three counters in the relevant brick. Choose a different colour for each brick.
- Show pupils how to push the counters from the bricks below into the brick above. Pupils can then subitise, count all or count on to find the sum to record in the brick. Ask them which strategy they used. They might also arrange the counters in dice patterns to help them remember how many in each group and support counting on from one of the groups. Say the number sentence together. Place another two counters into the middle brick so that pupils can add 2 and 3 in the same way to find the sum for the brick above. Say the number sentence together as before. Finally, push all counters together into the top brick. This will be too many to subitise. Ask pupils to organise the counters in a way that makes them easier to count. This could be in one long row or in dice patterns such as six to count on from. Remove the counters and use the numbers in the pyramid to say the final number bond, 3 + 5 = 8.
- Rearrange the numbers in the base and repeat.

Same-day enrichment

- Challenge pupils to repeat the pyramid activity with the numbers 2, 3 and 4. Ask them to predict how many different pyramids they will be able to make and which numbers will be at the top of the pyramid.
- Share solutions. Pupils should recognise that they can make the same number of different pyramids as they did with 1, 2 and 3. (Although the numbers have been changed, there are still only three different numbers they can use.)

Challenge and extension question

Question 4

4 Use the picture to write different addition sentences.

☐ + ☐ = ☐   _____
☐ + ☐ = ☐   _____
☐ + ☐ = ☐   _____

This question requires pupils to find as many different addition sentences as they can. No examples are provided for support, though pupils can look back to previous questions if they need to see how to set out number sentences.

Pupils might link groups according to type of animal, or in other ways. Encourage pupils to compare their addition sentences with another pupil to check their sums and to see if they have missed any. Some pupils may challenge themselves by adding sheep and chickens or some other combination that extends beyond 10. Support their calculations and recording of the numbers beyond 10 if necessary.

Chapter 2 Addition and subtraction within 10

Unit 2.4
Addition (3)

Conceptual context

This is the fourth in a series of units on addition and subtraction within 10. This unit looks at further ways to record part–whole relationships as addition. The unit begins with telling a story, which is then recorded as an addition sentence. This helps to give real-life meaning to the process of addition, which helps pupils to communicate about addition in writing. The next part of the unit explores other formats for recording and representing addition. The arrow format gives a direction to the addition that can later be reversed and 'undone' as subtraction. This is a very 'light-touch' introduction to subtraction and its written sign, which will be explored in more fully from Unit 2.6.

Question 3 gives pupils more practice in finding sums, focusing on the strategy of counting on. This is a more efficient strategy than counting all. It is important that pupils explore how they count on and what support they use to tell them when to stop. This raises awareness of the ways in which errors could occur and gives pupils checking strategies. Sharing methods will help pupils identify methods they can relate to and use either initially or for checking. The challenge and extension question introduces using a symbol to represent a number, within the familiar format of an addition sentence. This activity encourages pupils to use what they know about addition sentences to reason about numbers.

The focus of the unit is on further reducing the support for writing addition sentences and introducing new formats without losing sight of what an addition sentence actually means. It moves on to encourage counting on to find a sum. The unit begins to prepare pupils for subtraction by introducing the knowledge that addition can be reversed or 'undone'; however, subtraction is not developed any further at this point.

It is important that, through these activities and questions, pupils have the opportunities to learn that:
- part–whole relationships can be recorded in a variety of ways
- addition can be undone or reversed. This is called subtraction and is recorded with –.

Learning pupils will have achieved at the end of the unit

- Pupils will have been introduced to the idea that an addition sentence records a number story (Q1)
- Pupils will have explored number stories in words and recorded them as addition number sentences (Q1)
- Pupils will have recorded addition sentences with a further reduced writing frame (Q1)
- Pupils will have explored a mapping format for recording addition sentences (Q2)
- Pupils will have been introduced to subtraction as the 'undoing' of addition and to the symbol – for recording subtraction (Q2)
- Pupils will have consolidated the use of efficient strategies for adding by counting on from the highest addend (Q3)
- Pupils will have further consolidated recording part–whole relationships using symbols in a number sentence (Q3)

Resources

mini whiteboards; interlocking cubes; model cars or animals; counting objects; 0–12 number cards; numeral and spot dice; 0–10 number lines

Vocabulary

add, addition, combine, addition sentence, story, addend, equals, sum, square, circle, circular, input, output, rule, counting on, pattern

Chapter 2 Addition and subtraction within 10

Unit 2.4   Practice Book 1A, pages 46–48

Question 1

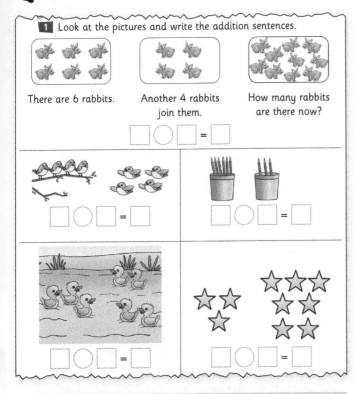

What learning will pupils have achieved at the conclusion of Question 1?

- Pupils will have been introduced to the idea that an addition sentence records a number story.
- Pupils will have explored number stories in words and recorded them as addition number sentences.

Activities for whole-class instruction

- Explain that an addition sentence tells a story. It starts off with a quantity of things, an addend in the addition sentence. Then more things need to be added – another addend. So, there are two addends. The addition sign + tells us that we need to put both quantities together and the equals sign = tells us that the number that comes after the equals sign, the sum, is how many there are altogether. Addends can be added in any order, the sum will still be the same.

- Following the interests of pupils or linking to a current topic, choose some things to use to tell an addition story, for example model cars (or magnetic models on a whiteboard). Modelling the story and recording it in a number sentence helps pupils to link the word sentence and the number sentence. Begin the story by drawing or showing a rectangular boundary for the whole car park.

- Then model five red cars arriving and parking: *There were five cars in the car park.* Draw a box around the five cars.

Continue the story with another three blue cars arriving to park in the car park. Draw a box around the three cars. Say: *There are eight cars in the car park altogether.* Point to the boundary of the whole car park. Show pupils the layout □○□=□ for an addition sentence and ask them which part of the word sentence belongs in each box. Confirm that the square boxes should contain the numbers and the circular box the addition sign. Ask pupils which numbers are parts and which number is the whole. Complete the number sentence for the cars in the car park: 5 + 3 = 8.

- Move on to use cubes, deciding that blue cubes should represent the blue cars and red cubes should represent the red cars. Set up five red cubes add two blue cubes. Can pupils explain what this represents?

- Let cubes represent pieces of fruit. Pupils decide colours for two pieces of fruit and represent, concretely, addition sentences that you write on board. Ask pupils to make up their own addition sentences with a sum up to 10 written using the layout □○□=□ and model their sentence using cubes. They could tell a partner what they represent, or the partner could 'read' the sentence and tell the story of what it represents.

- Complete Question 1 in the Practice Book.

Same-day intervention

- Instead of using cubes to model an addition number sentence, use equipment such as model cars, farm or zoo animals or something else of interest to pupils.

- Make two large versions of the layout □○□=□. Label the boxes in both versions: addend, add, addend, equals, sum. Ask pupils to place two objects in the first square, and write 2 in the matching position in the second layout. Confirm that the circular box is the place for the addition sign and ask a pupil to write it in. Agree how many to add, for example 4. Ask a pupil to place four objects in the second square box and write the number in the matching square box.

- Pupils work together to find the sum. They might count all, count on or recall a fact that they have regularly encountered. Say the completed addition story together first and then the matching addition number sentence.

- Repeat, either with pupils working in pairs or as a group.

Chapter 2 Addition and subtraction within 10

Unit 2.4 Practice Book 1A, pages 46–48

Same-day enrichment

- Give pupils sets of three numbers to make an addition story and number sentence from. Sets could include 3, 5 and 8; 9, 1 and 10; 6, 3 and 9, or something similar. Challenge pupils to write the matching number sentences and to record a story in both pictures and words for one of their addition sentences.

Question 2

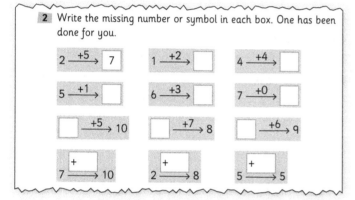

What learning will pupils have achieved at the conclusion of Question 2?

- Pupils will have explored a mapping format for recording addition sentences.
- Pupils will have been introduced to subtraction as the 'undoing' of addition and to the symbol ' – ' for recording subtraction.

Activities for whole-class instruction

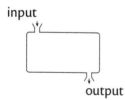

- Draw a simple function machine. Explain that when a number goes into the machine, something happens to it.
- Explain that 'input' is where you put something into the machine. The number travels through the machine where it is changed in some way and the result comes out of the output. Draw a table to show the results of putting 1 and 2 into the machine. Ask pupils to predict the output for an input of 3.

Input	Output
1	2
2	3
3	

- Ask pupils to explain what the machine did to each number. Agree that the machine added 1 to each number and record this on the front of the machine as:

$$\xrightarrow{+1}$$

- Explain that adding 1 is known as 'the rule' and is what happens to every number that goes into the machine. Call out an input for pupils to call back an output. Go on to explain that the rule can change. Give pupils an input and output and ask them to identify the rule. Switch to giving an input and a rule, asking pupils to give the output.
- Show pupils a quick way of recording an input, rule, output as shown in Question 2 in the Practice Book:

$$2 \xrightarrow{+5} (7).$$

Call out some further input/outputs or input/rules for pupils to work out and record in this way.

- Return to the function machine. Explain that some machines let you put the output back in through the output. When the number travels back through the machine, it undoes what it did before. You might need a discussion about undoing in everyday contexts (for example buttons, zips and melting ice cubes). So if the rule was +2 when a number went in to the machine, if you put the output back in, it will take away the 2 and you will be left with what you started with. Show pupils how this can be recorded, introducing the subtraction sign, as the sign that 'undoes' addition.

- Give pupils some inputs such as 1, 3, 5 and 7 and the rule +3. Ask them to record what happens when the numbers go through the machine and come back through the output in the format shown.

Chapter 2 Addition and subtraction within 10

Unit 2.4 Practice Book 1A, pages 46–48

Same-day intervention

- Draw a simple function machine on a mini whiteboard or use a large empty box on the desk. Place two objects at the input of the machine and explain that they are going to pass through the machine and another one will be added to them. Push the objects into the machine, add another object and push the objects through the output. Ask pupils to say what happened, for example: *Two objects went into the machine. The machine added one more object so three objects came out of the machine.* Agree the matching number sentence together, 2 + 1 = 3. Ask each pupil to take a turn at pushing objects into the machine, adding 1 and pushing them out of the machine. Ask the group to say the number story and write the matching number sentence together.

- Repeat the activity, but change the rule to adding 2. Give pupils another turn at putting some objects into the machine, adding to the quantity and pushing the objects out of the machine. Introduce or revisit the method of recording shown in the Practice Book:

  2 $\xrightarrow{+5}$ 7 .

- Use the same approach to explore how numbers can also come back through the machine, recording the whole process in the format:

  Push an object through the machine, adding 2, then send the same objects back the other way, removing 2. Record the addition sentence alongside the machine, then link to the shortened format to show both journeys through the machine.

Same-day enrichment

- Give pupils a set of number cards (0–12) to shuffle and turn over the top two cards. These two numbers are the input and output of a machine. Pupils record what the machine did using the double arrow format.
- After completing five machines, pupils collect and shuffle the cards to generate a further five pairs of numbers and machines.

Question 3

3 Write the answer in each box.

6 + 2 = ☐	3 + 5 = ☐	4 + 4 = ☐
5 + 0 = ☐	7 + 1 = ☐	5 + 4 = ☐
2 + 6 = ☐	3 + 2 = ☐	0 + 10 = ☐
7 + 3 = ☐	3 + 3 = ☐	0 + 4 = ☐
1 + 9 = ☐	6 + 3 = ☐	
0 + 8 = ☐	2 + 4 = ☐	

What learning will pupils have achieved at the conclusion of Question 3?

- Pupils will have consolidated the use of efficient strategies for adding by counting on from the highest addend.
- Pupils will have further consolidated recording part–whole relationships using symbols in a number sentence.

Activities for whole-class instruction

- Explain that counting is a good way to find a sum, but it does take time to count everything and it can be easy to make a mistake. So, it is better to reduce the amount of counting by starting the count from one of the numbers given rather than from zero. Starting from the higher number reduces the counting further.

- Use stickers to replace the 6 on spot and numeral dice with 0 (so that totals will not exceed 10). Roll one of each dice and find the sum by counting on from the number shown on the numeral dice. Record the matching addition sentence.

- Discuss different ways to keep track of how many to count on. This might include double counting, that is counting on five by saying the next five numbers while simultaneously raising 1 to 5 fingers, counting on 1 more for each spot shown on the dice or something else.

- Give pairs of pupils a numeral dice and a spot dice to roll to generate two addends. Pupils record their addition sentence and find the sum by counting on from the numeral shown on the dice.

- After a few minutes, ask pupils to stop and look at their sums. Which sum (total) have they recorded more often? Do all pupils agree? They might be able to explain why, for

Chapter 2 Addition and subtraction within 10 Unit 2.4   Practice Book 1A, pages 46–48

example: with three ways to make a sum of 4, 5 or 6, these totals are likely to appear more frequently than 2, 3, 7 and 8 with two ways and 0, 1, 9 and 10 with just one way.
- Complete Question 3 in the Practice Book.

Same-day intervention

- Show pupils a 0–10 number line. Count along it together. Place a counter next to each number, counting along the number line each time another counter is added. Remove the counters.
- Give pairs of pupils a numeral dice and a spot dice to roll to generate two addends. Pupils place a finger on the number on the number line matching the first addend. They take the same number of counters as the second addend. Pupils use the counters as markers on the number line, so if 5 and 3 were rolled, with a finger on 5 to show which number to start from, the three counters are placed next to 6, 7 and 8. The last counter gives the sum. Pupils should then say the numbers with counters next to them, to help to develop counting on. Pupils record the addition sentence 5 + 3 = 8.
- Check that pupils can see how the numbers in the addition sentence are linked to those on the number line.
- Repeat, rolling the dice to generate new addends.

Same-day enrichment

- Choose three sums from Question 3 for pupils to use as the first addition sentence in a pattern of calculations. For example, 0 + 4 = 4 could be followed by 0 + 5 = 5, 0 + 6 = 6, 0 + 7 = 7 … Ask pupils to say what they notice about each pattern, for example: *The sum is the same as the number added to 0.*

Challenge and extension question

Question 4

4 Think carefully and then write the correct number in each box.

(a) If ★ + ✿ = 8 and 5 + ✿ = 7, then

★ = ☐ and ✿ = ☐

(b) If ◉ = ★★★★ and ★ = △△, then

◉ = ☐

This question requires pupils to use what they know about addition sentences to work out what a symbol within a sentence represents. They can then substitute the value of that symbol into another addition sentence to work out the value of a different symbol. The process is the same in part (b), but now the relationships are described entirely with symbols in addition sentence format. Pupils need to relate these two sentences to find the value of the ring. Both (a) and (b) require a number of steps to be completed to reach a solution. Encourage pupils to record the value of a particular symbol above that symbol. As they find the value of the flower and record it above or next to each image of the flower, they are already substituting its value into the next addition sentence.

Pupils may not realise that the value of the flower and star are consistent throughout part (a) but there is no link with part (b) where the value of the star is not fixed. Part (b) requires logical and algebraic thinking (something that young pupils are indeed capable of) and what is important is that pupils recognise that the ring is equivalent to six triangles – the star is simply a step along the way. Pupils do not need to find a numeric value for any of the symbols in part (b).

Chapter 2 Addition and subtraction within 10

Unit 2.5
Let's talk and calculate (1)

Conceptual context

This is the fifth in a series of units on addition within 10. This unit looks at addition in a variety of different representations. The next unit will move on to consider subtraction.

In essence, this is a summary unit before moving on to subtraction, providing opportunities for practice and consolidation to develop fluency.

Learning pupils will have achieved at the end of the unit

- Pupils will have reinforced the link between concrete objects and abstract symbols through modelling and recording addition number sentences (Q1)
- Pupils will have explored number stories presented in a variety of representations and recorded them as addition number sentences (Q1)
- Knowledge of the commutative law for addition will have been further reinforced (Q1, Q2, Q3)
- Pupils will have sorted addition sentences according to their sum (Q2)
- Pupils will have worked systematically to find all possibilities and will be able to explain how they know that they have found all possible solutions (Q2)
- Fluency with number bonds to 10 will have been further developed (Q2)
- Pupils will have had further practice of finding the sum in addition number sentences (Q3)
- Pupils will have explored finding an addition number sentence to match a sum (Q3)
- Pupils will again have been encouraged to recognise efficient strategies for adding by counting on from the highest addend (Q3)

Resources

bananas; ten-frames; Hungarian 'number pictures'; counting objects for example model cars; sets of 4–10 number cards; dice; 0–10 number line

Vocabulary

part, whole, addend, sum, add, addition, number sentence, ten-frame, count on, number line

Chapter 2 Addition and subtraction within 10

Unit 2.5   Practice Book 1A, pages 49–51

Question 1

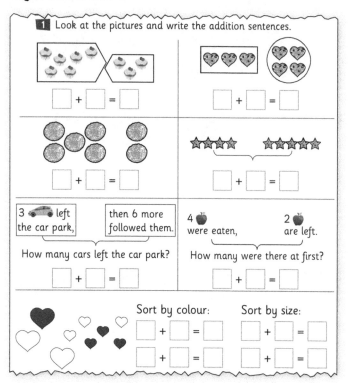

What learning will pupils have achieved at the conclusion of Question 1?

- Pupils will have reinforced the link between concrete objects and abstract symbols through modelling and recording addition number sentences.
- Pupils will have explored number stories presented in a variety of representations and recorded them as addition number sentences.
- Knowledge of the commutative law for addition will have been further reinforced.

Activities for whole-class instruction

- Show pupils a bunch of nine bananas and explain that this is the whole bunch of bananas. Count the bananas together and record the total. Explain that you are going to split the bunch into two parts, so that you can keep some bananas in school and take some home. Ask pupils to suggest how you should split the bunch and why.

- After a few minutes' discussion, give pupils a piece of paper and ask them to record their suggested parts and the whole in a number sentence, with their name on the reverse. Collect all the number sentences and sort them. Choose the most or least frequently occurring number sentence as the winner. Without telling pupils what the winning number sentence is, invite one pupil who suggested the winning number sentence to come and split the bunch of bananas to match their number sentence. Reiterate the language of parts and whole repeatedly.

- Ask pupils what the number sentence must have been. Confirm that the two addends (addends are the same as parts) could occur in any order and reveal both versions of the winning number sentence. Check the number sentence against the arrangement of the bananas and then swap over the two bunches to check the second version. Tell pupils that they have helped a little, but which part should you take home and which part should you leave at school? Agree, say: *6 bananas for school and 3 for home, 9 bananas in total, 6 + 3 = 9* to prepare pupils for the wording used in the Practice Book.

Same-day intervention

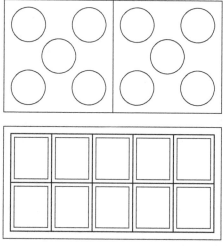

- Give pupils a Hungarian 'number picture' and a standard ten-frame. Count the number of spaces on each frame to confirm that they both have ten spaces. Ask pupils to explain how they are the same and how they are different. Explain that the Hungarian number picture is particularly useful for subitising quantities. Using one half of the frame, ask pupils to show you where they could place 1, 2, 3, 4 and 5 counters so that they can subitise them. Provide pupils with a dice for comparison of layout.

- Check that pupils recognise that one half of the Hungarian number picture is 5, so 5 add 5 equals 10, 5 + 5 = 10.

- Ask pupils to place five counters of one colour on the left-hand side of the Hungarian number picture and one counter on the other side of the ten-frame.

Chapter 2 Addition and subtraction within 10

Unit 2.5 Practice Book 1A, pages 49–51

Model counting on from 5 by saying 5 + 1 = 6, pointing to the relevant parts of the frame.

- Remind pupils that they know there are five counters in the left-hand side of the ten-frame and cover them with a piece of card. Repeat the number sentence: 5 + 1 = 6, pointing to the covered counters and the single counter.
- Ask pupils to add a second counter to the right-hand side of their Hungarian number picture. Give each pupil a piece of card to cover the left-hand side of the frame. Support them to count on from 5 to find the sum, 7, and to say the number sentence to match their frame, pointing to the relevant part, 5 + 2 = 7.
- Continue with 5 + 3, 5 + 4 and 5 + 5, reinforcing counting on from 5 each time. Allow pupils to lift the card to check there are five counters if they need to remind themselves.

Same-day enrichment

- Challenge pupils to draw or arrange objects so that the layout can be recorded using two different pairs of addition number sentences.
- If necessary, remind them of the last part of Question 1 in the Practice Book.
- Pupils could take turns to challenge each other to record both pairs of addition number sentences for each other's arrangements.

Question 2

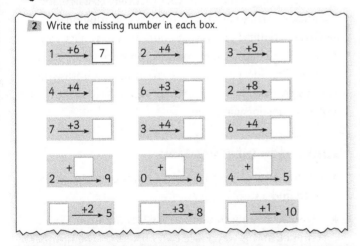

What learning will pupils have achieved at the conclusion of Question 2?

- Pupils will have further explored a mapping format for recording addition sentences.
- Pupils will have sorted addition sentences according to their sum.
- Pupils will have worked systematically to find all possibilities and will be able to explain how they know that they have found all possible solutions.
- Knowledge of the commutative law for addition will have been further reinforced.
- Fluency with number bonds to 10 will have been further developed.

Activities for whole-class instruction

- Ask pupils to look at Question 2 in the Practice Book and remind you what the format of the questions means. Ask pupils to identify parts and wholes, addends and sums for some of the earlier and later parts of the question. Ask pupils to count how many of the questions are asking them to find the sum, how many are asking them to find the first addend and how many times they need to find the second addend.
- Tell pupils that you would like them to answer the whole of Question 2 and then decide which sum occurs most often. After identifying which number sentences for that sum are included in Question 2, they should record the rest of the possible addition number sentences for that number.
- Check that pupils recognised that the sum 10 occurs most frequently, with 5 + 5 = 10, 0 + 10 = 10 and 10 + 0 = 10 as the other possible addition number sentences. Pupils should also have suggested reversing the order of the addends in the addition number sentences for 10 included within Question 2: 8 + 2 = 10, 3 + 7 = 10, 4 + 6 = 10 and 1 + 9 = 10.

Same-day intervention

- Make a table with six columns headed 5, 6, 7, 8, 9 and 10 on a large piece of paper. As each calculation is explored, identify the sum and record the calculation in the correct part of the table.
- Work through the addition mapping statements in Question 2, using counting objects to model each addition and find the sum. Count out the objects to represent the first addend and place them on a number line, counting along to check that the last number is the same as the written addend. Then count out the objects representing the second addend and continue to place these along the number line. Count on from the objects representing the first addend to find the sum, confirming that this is the last number with an object next to it on the number line.

Chapter 2 Addition and subtraction within 10

Unit 2.5   Practice Book 1A, pages 49–51

- Where an addend is missing, count out objects to represent the known addend and place these on the number line. Point out the known sum and explore how many more objects must be placed on the number line to get there. Count these objects to find the missing addend.
- When all calculations have been completed, identify which column has the most calculations in it. Alongside each calculation in this column, swap over the addends. Count from 1 to 10 to identify which numbers have not been used as an addend (0, 5 and 10). Using the same approach of placing the number of objects matching the known addend on the number line, identify the second addend for each calculation and record them in the column. Pupils should recognise that 0 + 10 = 10 and 10 + 0 = 10 have now been recorded in the column, so there is no need to swap the addends over in these calculations. Ask pupils to tell you why there is no need to swap the addends in 5 + 5 = 10.

Same-day enrichment

- Say to pupils that you heard someone say that to write down all the addition number sentences for 10, you need to write each number from 0 to 10 twice. Ask pupils if they think this is true. Challenge them to prove their response in some way. If pupils need help to get started, suggest that they write out all the addition number sentences for 10.
- It is true that the numbers 0 to 10 are written twice as addends, but 10 is also written 11 times as the sum, so 10 is recorded 13 times in total.

Question 3

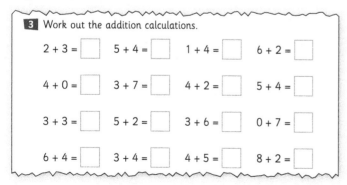

What learning will pupils have achieved at the conclusion of Question 3?

- Pupils will have had further practice of finding the sum in addition number sentences.
- Pupils will have explored finding an addition number sentence to match a sum.
- Pupils will again have been encouraged to recognise efficient strategies for adding by counting on from the highest addend.
- Knowledge of the commutative law for addition will have been further reinforced.

Activities for whole-class instruction

- Challenge pupils to complete each number sentence in Question 3 as part of a game.
- Give pairs of pupils a set of number cards with the number 4–10 on. Pupils shuffle the cards, place the pile face down and take turns to turn over the top card. If their card has a number that is the sum that completes of one of the number sentences in Question 3, they should put a tick next to that sentence in their book. Pupils take turns to turn over a card. Once all the cards have been used, pupils shuffle the cards again and continue in the same way. If there is no number sentence with a sum matching the number card, the pupil must miss that turn. The first pupil to tick all the addition number sentences is the winner.
- Completing the addition number sentences as part of a game encourages pupils to work mentally to calculate as quickly as they can. This promotes counting on. If necessary, remind pupils that counting on is a fast way to add, particularly when counting on from the highest addend. Provide a number line to support counting on. Games give pupils the opportunity to practise a particular skill in a fun, engaging way.

Same-day intervention

- Remind pupils how they used a number line to count on to find a sum.
- Revisit by finding the sum of the first addition sentence in this way. Place two objects on the number line, then a further three. Count on from 2, saying the next counting number for each object representing the second addend. Confirm that the sum is reached when all have been counted.

Chapter 2 Addition and subtraction within 10

Unit 2.5   Practice Book 1A, pages 49–51

- Leaving the objects on the number line, show pupils how to place a finger on the first addend on the number line (2), then use a finger on the other hand to count on the second addend (3). The final number their finger landed on is the sum. Explain that this is the same as placing objects on the number line, but quicker because there is no need to match the objects to the number line.
- Find the sum of 5 + 4 using objects on the number line and then fingers. Encourage pupils to find the rest of the sums using fingers only.
- Once the sums have been found, pupils could play the game from the whole-class instruction section. With the sums already completed, the game will be easier to play and quicker.

Same-day enrichment

Sum	4	5	6	7	8	9	10
Number of times occurring	1	2	2	3	1	4	3

- Either show pupils the table of sums for Question 3 or draw up a table together.
- Ask pupils if they think the game played during the whole-class part of the session was fair. If necessary, point out that there was only one question with a sum of 4 or 8, so players may have had to wait a long time to turn over the 4 or 8 cards and complete that addition sentence.
- Give pupils some blank cards and suggest that they make extra number cards to make the game fair. Which number cards should they make?
- Alternatively, pupils could change the addition sentences so that each sum occurs the same number of times.

Challenge and extension question

Question 4

4 Write the two addition number sentences for each domino.

☐ + ☐ = ☐   ☐ + ☐ = ☐
☐ + ☐ = ☐   ☐ + ☐ = ☐
☐ + ☐ = ☐   ☐ + ☐ = ☐
☐ + ☐ = ☐   ☐ + ☐ = ☐
☐ + ☐ = ☐   ☐ + ☐ = ☐
☐ + ☐ = ☐   ☐ + ☐ = ☐
☐ + ☐ = ☐   ☐ + ☐ = ☐
☐ + ☐ = ☐   ☐ + ☐ = ☐
☐ + ☐ = ☐   ☐ + ☐ = ☐
☐ + ☐ = ☐   ☐ + ☐ = ☐

Which domino has only one addition number sentence?

This challenge and extension question asks pupils to record two addition number sentences for each domino. Encourage pupils to subitise the spot arrangement on each half of the domino. Since the dominos are pictured vertically, there is no indication of which number should be the first addend and which should be the second. The order does not matter, provided both are recorded. This will help to reinforce the commutative nature of addition. Pupils should recognise that they can only write one addition number sentence where both halves of the domino show the same number.

Chapter 2 Addition and subtraction within 10

Unit 2.6
Subtraction (1)

Conceptual context

This is the first in a series of four units introducing subtraction. In this unit, the focus is on subtraction as taking away. Although taking away is a useful introduction to subtraction for pupils because it is easy to understand, it is only one kind of subtraction. Taking away links to what pupils already know about parts and whole, and is the natural link as the inverse of addition. Pupils will quickly move on to explore subtraction as 'comparing to find the difference', which is the approach generally used in everyday life, to build a deeper, stable understanding of subtraction.

This is explored using a range of physical resources so that pupils experience taking away as the removal of some objects. The language of subtraction is introduced and compared with that of addition. Pupils are also reminded that addition is commutative and discover that subtraction is not.

The focus of the unit is on introducing subtraction in a similar format to how addition was introduced. This helps to alert pupils to the fact that addition and subtraction are closely related, an idea that will be further developed throughout this sequence of sessions.

It is important that, through these activities and questions, pupils have the opportunities to learn that:

- part–whole relationships can be recorded in a variety of different representations
- taking away is one form of subtraction
- subtraction is not commutative.

Learning pupils will have achieved at the end of the unit

- Pupils will have explored subtraction as taking away (Q1, Q2, Q3)
- Pupils will have been introduced to the idea that a subtraction number sentence records a subtraction number story (Q1)
- Pupils will have recorded part–whole relationships presented as images as number sentences (Q1)
- Pupils will have practised recording subtraction using number sentences (Q1, Q2)
- Pupils will have been introduced to the language of subtraction: minuend for the whole, the number being subtracted from; subtrahend for the part being subtracted; and difference for the part left after subtraction (Q3)
- Pupils will have explored recording part–whole relationships in an abstract format within a table, reinforcing the link between parts and wholes through recording in columns (Q3)
- Pupils have been introduced to the idea that although addition is commutative, subtraction is not (Q3)

Resources

pens; pen pots; cloth mats; counting objects; counters; double-sided counters; ten-frames; small world resources (e.g. animals, vehicles, people)

Vocabulary

subtraction, take away, part, whole, number sentence, number story, minuend, subtrahend, difference

Chapter 2 Addition and subtraction within 10

Unit 2.6 Practice Book 1A, pages 52–54

Question 1

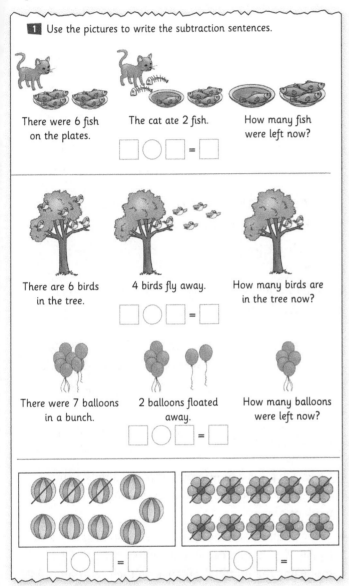

What learning will pupils have achieved at the conclusion of Question 1?

- Pupils will have explored subtraction as taking away.
- Pupils will have been introduced to the idea that a subtraction number sentence records a subtraction number story.
- Pupils will have recorded part–whole relationships presented as images in number sentences.
- Pupils will have practised recording subtraction using number sentences.

Activities for whole-class instruction

- Show pupils nine coloured pens. Count the pens together and place them in a pot. Take three pens out of the pot and put them on another table, still within sight. Together, count how many are left in the pot: 6.
- Ask pupils to tell a partner the story of what happened. Ask two or three pairs to share their story. Use these as a basis to agree the story for what happened: There were nine pens in a pot, three of the pens were removed from the pot, so there were six pens left in the pot.
- Explain that the pens taken away were part of the whole – all point to that 'part', the pens left were another part of the whole – all point to that 'part'. The two parts make up the whole – put the parts together again in the pot to make the whole that you began with. So the story is about a part–whole relationship.
- Show pupils the outline of the number sentence. Remind them that a part–whole relationship can be expressed in a number sentence. The squares contain numbers and the circular box is for the operation sign. They have already used + in an addition number sentence. Remind them that they used − as the sign for undoing addition in Unit 2.4.
- Explain that − means subtraction and that take away is one way of subtracting, so this is the operation sign to use in a subtraction number sentence. Complete the number sentence as you re-enact the story: $9 - 3 = 6$. This time there is no need to put the two parts back together again unless you feel the part–whole relationship needs reinforcing.
- Give each pair of pupils a pot of pens. Tell a take away story with pupils carrying out the story, then agree the number sentence to match the number story. Repeat two or three times until pupils are confident at recording the number sentence.
- Move on to drawing eight balloons. Ask pupils to start the number story, for example there were eight balloons. Begin to record the number sentence: 8 ◯ ☐ = ☐.
- Ask pupils to suggest what might happen to some of the balloons and agree the next part of the story, for example: *Two balloons popped*. Ask pupils to suggest how to record this in the number sentence, $8 - 2 = $ ☐.
- Cross out two of the balloons and ask pupils how to finish the story and complete the number sentence, for example: *There were six balloons left*, $8 - 2 = 6$.
- Ask pupils to complete Question 1.

Chapter 2 Addition and subtraction within 10   Unit 2.6 Practice Book 1A, pages 52–54

Same-day intervention

- Give each pupil a square of cloth to use as a mat, about 30 cm squared. Provide some objects that reflect the pupil's interests such as toy cars, model animals or play people.
- Ask pupils to each choose five animals to place on their mat. Start off the story with pupils repeating it, for example there were five animals in the field. Tell pupils that two animals ran off through a hole in the fence. Ask them to remove two of their animals, placing them under the corner of the mat so they are not visible but pupils are aware that they are still there. Ask pupils to tell you how many animals are left in the field. (3) Ask pupils to return their animals to the field and retell the story, moving the animals accordingly. There were five animals in the field. Two ran off through a hole in the fence so there were three animals left in the field. Remind pupils that 2 is a part, 3 is a part and 5 is the whole. Show pupils the outline number sentence and retell the story, completing the number sentence, 5 (whole) – 2 (part) = 3 (part).
- Ask pupils to return the animals to the mat. Repeat with four of the five animals running away. Remind pupils that 4 is a part, 1 is a part, 5 is the whole. Show pupils the outline number sentence again and retell the story. Ask them to tell you how to complete the number sentence, starting with the whole: 5 – 4 = 1.
- Starting with a different number of animals, create two more subtraction stories, writing the matching subtraction number sentence together.

Same-day enrichment

- Give pupils three subtraction number sentences and ask them to model a matching story with counters, cubes or small counting objects. Remind them that they need to show the whole and parts each time. Pupils could draw a box around the objects and include an arrow to show part of the collection being removed or show taking away in some other way. Ask pupils to take turns telling and modelling their stories to a partner.

Question 2

> 2 Complete the subtraction calculations.
>
> 8 – 4 = ☐   10 – 9 = ☐   6 – 4 = ☐   2 – 2 = ☐
>
> 4 – 3 = ☐   5 – 2 = ☐   9 – 4 = ☐   6 – 1 = ☐
>
> 10 – 5 = ☐   8 – 7 = ☐   7 – 5 = ☐   8 – 6 = ☐
>
> 5 – 5 = ☐   2 – 0 = ☐   4 – 1 = ☐   9 – 6 = ☐

What learning will pupils have achieved at the conclusion of Question 2?

- Pupils will have continued to explore subtraction as take away.
- Pupils will have practised recording subtraction using number sentences.

Activities for whole-class instruction

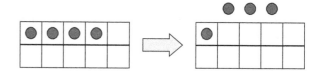

- Give each pupil a ten-frame and ten counters. Explain that they are going to use the ten-frame as their working area.
- Write a subtraction number sentence such as 4 – 3 = ☐. Explain that pupils can find the missing number by modelling the number sentence on their ten-frame. Ask pupils to place four counters on the ten-frame. Remind them that it is easier to check how many they have on the ten-frame if they place the counters together, for example in a row.
- Highlight the next part of the subtraction number sentence, – 3 and ask pupils what this means. Agree that it means subtract three counters from the four counters, so we should take three counters away from the 4. Confirm that there is one counter left on the ten-frame and complete the subtraction number sentence, 4 – 3 = 1.
- Repeat with 8 – 4 = ☐. This is the first of the subtraction number sentences pupils are asked to calculate in Question 2. Pupils can use the ten-frame and counters to help them complete the calculations.

Chapter 2 Addition and subtraction within 10

Unit 2.6 Practice Book 1A, pages 52–54

Same-day intervention

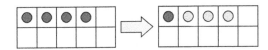

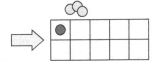

- Give each pupil a ten-frame and ten double-sided counters. Explain that they are going to use the ten-frame as their working area.

- Write a subtraction number sentence such as 4 – 3 = ☐. Explain that pupils are going to find the missing number by modelling the number sentence on their ten-frame. Ask pupils to place four counters on the ten-frame, all showing the same colour. Remind them that it is easier to check how many they have on the ten-frame if they place the counters together, for example all in a row.

- Read along the number sentence, confirming that it says 4 take away 3. Ask pupils to turn over three of the counters so they show the other colour. Point out that there are still four counters on the ten-frame; they have not carried out the next part of the number sentence yet. What they have done is get ready. Remind them that the number sentence says take away 3 and they now have three counters ready to take away from the rest. These now show a different colour. Ask pupils to predict how many counters will be left on the ten-frame when they have taken away the three differently coloured ones.

- Ask pupils to remove the three counters from the ten-frame, counting them off one at a time. Confirm that they had four counters, took three away and there is now one counter left on the ten-frame. Complete the subtraction number sentence, 4 – 3 = 1.

- Repeat with 5 – 2 = ☐, taking care to use the same step-by-step approach.

Same-day enrichment

- Ask pupils to look at the calculation 2 – 0 = from Question 2. Check if pupils all had the same solution, 2.
- Confirm that the starting number was unchanged by subtracting 0. Ask pupils if this will always be true. Challenge pupils to explain or demonstrate their conclusion in some way.

Question 3

> **3** Write the missing numbers in the spaces.
> Remember: minuend – subtrahend = difference

minuend	8	10	5	9	3
subtrahend	2	4	3	1	3
difference					

minuend	7	5	4	2	6
subtrahend	1	5	3	1	4
difference					

What learning will pupils have achieved at the conclusion of Question 3?

- Pupils will have further explored subtraction as taking away.
- Pupils will have practised recording subtraction using number sentences.
- Pupils will have been introduced to the language of subtraction: minuend for the whole, the number being subtracted from; subtrahend for the part being subtracted; and difference for the part left after subtraction.
- Pupils will have explored recording part–whole relationships in an abstract format within a table, reinforcing the link between parts and wholes through recording in columns.
- Pupils will have been introduced to the idea that although addition is commutative, subtraction is not.

Activities for whole-class instruction

- Write a subtraction number sentence such as 7 – 5 = 2. Remind pupils that to help us to identify the parts of a number sentence, each part has a name. For addition, this was addend + addend = sum. Remind pupils that the addends are the parts in a part–whole relationship. Both parts are called addends and they can be written in any order, the total would be the same: 3 + 4 = 7 and 4 + 3 = 7. The sum is how many altogether, the whole.

- For subtraction, the correct names for the whole and parts are: minuend – subtrahend = difference. Ask pupils what they notice. Check that pupils notice that all the names are different; this means they cannot be written in any order.

Chapter 2 Addition and subtraction within 10

Unit 2.6 Practice Book 1A, pages 52–54

- Give pupils a ten-frame and some counters and ask them to model 7 – 5 on the ten-frame. Check that the difference is 2, 7 – 5 = 2. Ask pupils to place five counters on their ten-frame and subtract 7. They should recognise that it cannot be done. Record this next to the original subtraction number sentence to reinforce that it is nonsense to swap over the minuend and the subtrahend. 7 – 5 = 2 but 5 – 7 ≠ 2.
- So, for 7 – 5 = 2, 7 is the minuend, the number that will be subtracted from, the whole. 5 is the subtrahend, the number that is taken away from the minuend. 2 is the difference, how many are left after subtraction.
- Ask pupils to model a minuend of 6 on their ten-frame. Tell them that the subtrahend is 5 and ask them what this means. Confirm that they need to subtract 5. Ask pupils to record the matching subtraction number sentence: 6 – 5 = 1. Repeat for a minuend of 8 and a subtrahend of 5. Ask pupils to confirm the difference and record the subtraction number sentence. Ensure that pupils can see the sentence, minuend – subtrahend = difference, as they complete Question 3 with a ten-frame and counters for support.

Same-day intervention

- Give pupils a copy of the sentence: minuend – subtrahend = difference, and a subtraction number sentence such as 5 – 3 = 2.
- Ask pupils to take five cubes and join them together in a stick. Tell pupils that this is the minuend. The number sentence continues with 'take away 3'. Show pupils how to break three cubes off the five so that they have two left. This is the difference, the 2 in the number sentence.
- Give pupils a new number sentence such as 6 – 4 = ☐. Read the number sentence together: 6 take away 4 equals something. Ask pupils which number is the minuend. Agree that it is the first number, the number they will take away from. Ask pupils to make a stick of six cubes. Agree that the next part of the number sentence says 'take away 4' and ask pupils what they should do. Confirm that this means they should take 4 away from the 6. As pupils do this, ask them to confirm how many they have left. Remind them that this is called the difference and ask them to complete the number sentence.
- Repeat the activity with a few more number sentences, reinforcing the vocabulary at each step of the calculation.

Same-day enrichment

- Give pupils a copy of the minuend, subtrahend, difference table with the numbers 0 to 10 in the difference row, in order. Challenge pupils to complete the table.
- Ask pupils to compare tables. Have they completed the table in the same way? Are all tables correct? If pupils used a pattern of calculations in their table, challenge them to describe it.

Challenge and extension question

Question 4

4 Write subtraction calculations with a difference of 5.

☐ – ☐ = 5   ☐ – ☐ = 5   ☐ – ☐ = 5

☐ – ☐ = 5   ☐ – ☐ = 5   ☐ – ☐ = 5

Pupils are asked to find subtraction calculations with a difference of 5. Of course, there are an infinite number of solutions. If pupils find it hard to know where to start, suggest they begin with 5 – 0 = 5, and then consider how they can change this to ensure that the difference of 5 is maintained. For example, if they began with a whole of 6, taking away 1 would leave a difference of 5. Alternatively, pupils could glance back at Questions 2 and 3 to see if there are any calculations with a difference of 5. Question 2 has 10 – 5 and 9 – 4 so pupils may choose to create a pattern of calculations from one or both of these instead.

The language of subtraction

The whole is called the minuend, the number being subtracted from. The part being subtracted from the minuend is called the subtrahend. The difference is the remaining part after subtraction.

minuend – subtrahend = difference

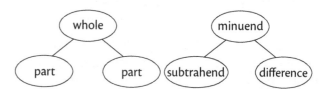

Chapter 2 Addition and subtraction within 10

Unit 2.7
Subtraction (2)

Conceptual context

This is the second in a series of four units introducing subtraction. The focus of the unit is to expand pupils' understanding of subtraction as taking away by using number bonds to find a missing part in a part–whole representation; pupils will learn that a missing 'part' is the 'difference'.

Bar models and partitioning trees are excellent images for promoting secure understanding of this concept. Partitioning trees are used within this unit and bar models will be revisited in Unit 2.9. Pupils' already developing knowledge of partitioning wholes into their constituent parts will support their ability to recognise these relationships and connections. These are also explored using a range of physical resources so that pupils experience these aspects of subtraction practically. The language of subtraction is reinforced and pupils continue to explore the link between addition and subtraction.

It is important that, through these activities and questions, pupils have the opportunities to learn that:

- part–whole relationships can be recorded in a variety of different representations
- taking away, difference and using linked addition number bonds are all forms of subtraction
- subtraction is not commutative.

Learning pupils will have achieved at the end of the unit

- Pupils will have revisited recording part–whole relationships in abstract formats such as partitioning trees (Q1)
- Pupils will have continued to explore subtraction as taking away (Q1, Q2, Q3, Q4)
- Pupils will have practised recording subtraction using number sentences (Q1, Q2, Q3, Q4)
- Pupils will have practised using the language of subtraction (Q1, Q2, Q3, Q4)
- Pupils will have recognised that partitioning trees can be used to record subtractions as well as additions (Q1)
- Pupils will have continued to explore the link between addition and subtraction, using addition bonds (Q1)
- Pupils will have begun to explore the family of facts that can be recorded from a part–whole relationship (Q1)
- Pupils will have explored a mapping format for recording subtraction (Q3)
- Pupils will have been introduced to the idea that addition 'undoes' subtraction and reminded that subtraction 'undoes' addition (Q3)
- Pupils will have explored subtraction as 'comparing to find the difference' (Q4)

Resources

mini whiteboards; counting objects, including interlocking cubes, counters and toy cars; function machine; subtraction pyramids; picture or photograph of ten cars; individual 0–20 number lines; wire coat hanger; pegs; small cloth; small world people

Vocabulary

partitioning tree, subtraction, take away, part, whole, number sentence, number story, minuend, subtrahend, difference, difference between, commutative, function machine, input, output, table

Chapter 2 Addition and subtraction within 10   Unit 2.7   Practice Book 1A, pages 55–56

Question 1

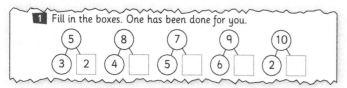

What learning will pupils have achieved at the conclusion of Question 1?

- Pupils will have revisited recording part–whole relationships in abstract formats such as partitioning trees.
- Pupils will have continued to explore subtraction as taking away.
- Pupils will have explored subtraction using addition bonds.
- Pupils will have practised recording subtraction using number sentences.
- Pupils will have practiced using the language of subtraction.
- Pupils will have recognised that partitioning trees can be used to record subtractions as well as additions.
- Pupils will have continued to explore the link between addition and subtraction.
- Pupils will have begun to explore the family of facts that can be recorded from a part–whole relationship.

Activities for whole-class instruction

- Draw a part–whole diagram for 5, 3 and 2 as above. Ask pupils to remind you what the diagram shows. Agree that 5 is the whole and that 3 and 2 are parts. Ask pupils to tell you the two addition number sentences for the diagram.
- Reinforce this by modelling it with a coat hanger and pegs.
- Now delete the 2 leaving an empty box in the part–whole diagram. Tell pupils that, although they know the value of the number in the box, you would like them to think about how they can find out the missing value when they don't know it. Continue to model this by covering two pegs on the coat hanger with a cloth.
- Agree that it could be a number fact that they are beginning to remember, 3 + 2 = 5, so the missing number must be 2. Ask what they could do if they don't remember the fact they need.

- If necessary, use five objects to remind pupils that 5 is the whole and 3 is one of the parts. So if they had five objects (the whole), they could take away 3 of them (a part) to find the other part, 2. Summarise this as a number sentence, 5 – 3 = 2. Model by revealing the two covered pegs on the coat hanger.
- Ask pupils to record three number facts they know in the partitioning tree format. Choose a fact to display and ask pupils to record the number sentences for that fact. Pupils are likely to record the two addition sentences but remind them that there are also two possible subtraction number sentences too.
- As pupils complete Question 1, ask them to think about how they are completing each partitioning tree. If they remember an addition fact, they should write + next to the partitioning tree. If they subtract the known part from the whole, that is the subtrahend from the minuend, they should write – next to the partitioning tree. This will help you to see which strategies pupils are using.

Same-day intervention

- Draw the outline of a partitioning tree on a piece of paper or mini whiteboard. Ask pupils to get five things and place them in the box at the top of the partitioning tree. Remind them that this is the whole: the minuend, when we are thinking about subtraction.
- Ask pupils to place three of their things in one of the part boxes and to confirm that they have two things left to go in the other part box. Pointing to the relevant boxes, show pupils that 5 take away 3 equals 2. Ask pupils to repeat the sentence, pointing to their own arrangements as they speak. Write the statement minuend – subtrahend = difference and record the number sentence 5 – 3 = 2 below it. Ask questions such as: Which number is the minuend (or subtrahend or difference)? Which number is the whole (or part or other part)?
- Repeat with a different whole and parts. Ask pupils to say and record the number sentence each time.
- Finally, draw a partitioning tree with a missing part and ask pupils to use the same method to identify the missing part. Ask pupils to say and record the number sentence, identifying the minuend, subtrahend and difference.

Chapter 2 Addition and subtraction within 10

Unit 2.7 Practice Book 1A, pages 55–56

Same-day enrichment

- When pupils have completed the partitioning trees in Question 1, ask them to record all the number sentences they can 'see' in each tree.
- Pupils should recognise the part–whole diagram and be able to write two addition number sentences since addition is commutative.
- Pupils may write one subtraction sentence echoing how they completed the partitioning tree. Prompt them to write another by suggesting that either of the parts could be the subtrahend.

Question 2

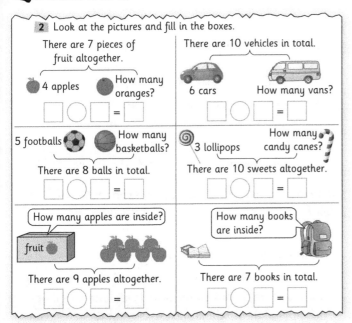

What learning will pupils have achieved at the conclusion of Question 2?

- Pupils will have continued to explore subtraction as taking away.
- Pupils will have continued to explore using a subtraction number sentence to record a subtraction number story.
- Pupils will have continued to practise using the language of subtraction.

Activities for whole-class instruction

- Remind pupils that they told some subtraction stories in the previous unit. Recap the story of the cat eating two of the fish in Question 1, Unit 2.6, and how this was recorded in a number sentence.

- Explain that stories can take a while to tell, so you are going to tell them some short car park stories and you would like them to record your stories in number sentence format.
- Using toy cars, model the stories as you say them, for example: *There were eight cars in the car park. Three of them left. How many cars are in the car park now?* Check that pupils have recorded 8 − 3 = 5. Repeat the story, identifying the minuend, subtrahend and difference.
- Make up another story with no cars leaving, for example: *There were nine cars in the car park. Four of them were white. How many cars were not white?* Check that pupils have recorded 9 − 4 = 5. Ask them to identify the minuend, subtrahend and difference in the number sentence and in your story.
- Tell some more stories about the cars until pupils can confidently capture the story in a subtraction number sentence. Make sure that stories focus on identifying the second part in the part–whole relationship but that the story does not always involve removing that part. This will help pupils to recognise when a subtraction calculation is appropriate in problem solving.

Same-day intervention

- Give each pupil some play people and act out some friends in the park stories.
- Begin with five friends in the park, the minuend. Three friends (subtrahend) leave to go home. Ask pupils how many friends are left, reminding them that this is the difference. Display the sentence minuend − subtrahend = difference and record the story as 5 − 3 = 2 alongside the written sentence.
- Ask pupils to put six friends in the park, using a piece of paper or fabric to represent the park. Ask one pupil to choose how many friends leave. Pupils model the action and record the number sentence as before.
- Ask pupils to put seven friends in their park. Look at the play people and ask pupils to imagine it is a hot, sunny day and three of the play people are wearing sunhats. Challenge pupils to say how many play people are not wearing sunhats and to say and record the subtraction number sentence. Pupils support each other so that everyone in the group has made up a park story and recorded it in subtraction number sentence format.

Chapter 2 Addition and subtraction within 10

Unit 2.7   Practice Book 1A, pages 55–56

Same-day enrichment

- Give pupils a collection of ten counting bears or similar. Ask them to sort them into two groups. They might choose to sort them by colour or size or some other attribute.
- Ask pupils to make up three bear stories, recording them in a similar way to the second part of Question 2, including the subtraction number sentence.

Question 3

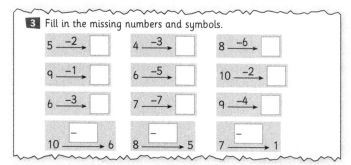

3 Fill in the missing numbers and symbols.

What learning will pupils have achieved at the conclusion of Question 3?

- Pupils will have continued to explore subtraction as taking away.
- Pupils will have explored a mapping format for recording subtraction.
- Pupils will have been introduced to the idea that addition 'undoes' subtraction and reminded that subtraction 'undoes' addition.
- Pupils will have continued to practise using the language of subtraction.

Activities for whole-class instruction

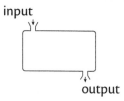

input

output

- Draw a simple function machine and remind pupils that they explored addition with a machine like this. Recap addition by exploring and completing the table:

Input	Output
4	7
5	8
6	

- Agree that the machine is adding 3 each time a number is put into it. Label the front of the machine with + 3 and ask pupils if they can remember how they recorded the additions with an arrow, for example: 4 $\xrightarrow{+3}$ 7. Record the other additions from the table in the same way.
- Draw a new machine and explain that this one subtracts. Label the front of the machine with – 3 and remind pupils that this is the rule that is what happens to every number that is put into the machine. Draw the table and ask pupils what the output would be for each input. Either extend the table or call out some other inputs for pupils to give you the output.

Input	Output
4	7
5	8
6	

- Record each calculation from the table with the arrow mapping format, 7 $\xrightarrow{-3}$ 4. Ask pupils what would happen if the output was fed back into the machine and came out of the input. Check that pupils understand that the subtraction would be 'undone', the amount taken away would be added back on. Show pupils how they could record this as 7 $\xrightleftharpoons[+3]{-3}$ 4. The arrows could be curved or straight. Check that pupils can see the number sentences 4 + 3 = 7 and 7 – 3 = 4 in the diagram.

Same-day intervention

- Draw a simple function machine on a mini whiteboard or use a large empty box on the desk. Place four objects at the input of the machine and explain that they are going to pass through the machine and one will be taken away from them. Push the objects into the machine, take away an object and push the rest of the objects through the output. Ask pupils to say what happened, for example: *Four objects went into the machine. The machine took one object away so three objects came out of the machine.* Agree the matching number sentence together, 4 – 1 = 3. Ask each pupil to take a turn at pushing objects into the machine, taking away one object and pushing the rest out of the machine. Ask the group to say the number story and write the matching number sentence together.
- Repeat the activity, but change the rule to taking away (subtracting) 2. Give pupils another turn at putting some objects into the machine, subtracting from the quantity and pushing the objects out of the machine. Introduce or revisit the method of recording shown in Question 3 in the Practice Book: 7 $\xrightarrow{-2}$ 5. Write

Chapter 2 Addition and subtraction within 10

Unit 2.7 Practice Book 1A, pages 55–56

- the subtraction number sentence alongside and compare. Remind pupils of the vocabulary of subtraction: minuend – subtrahend = difference. Ask pupils to identify each part of the number sentence in the arrow format.
- Use the same approach to explore how quantities can also come back through the machine. Push a number of objects through the machine, subtracting 2, then send the same objects back the other way, adding 2. Record the whole process in the same format:

$$7 \xrightarrow[+2]{-2} 5$$

- Record the addition number sentence alongside and check that pupils can see this in the arrow format.

Same-day enrichment

- Explain that two function machines can be joined together. The machines work in the same way as before. The output of the first machine becomes the input of the second machine. Something else happens to the input with the final result emerging from the second output.
- Ask pupils to design two connected machines where an input of 2 becomes a final output of 4. Challenge pupils to design a second set of connected machines that produces the same result but in a different way.

Question 4

4 Complete the subtraction calculations.

6 – 2 = ☐   4 – 4 = ☐   5 – 0 = ☐   7 – 2 = ☐
7 – 6 = ☐   9 – 4 = ☐   6 – 3 = ☐   3 – 2 = ☐
5 – 2 = ☐   7 – 3 = ☐   9 – 6 = ☐   10 – 5 = ☐
9 – 1 = ☐   6 – 5 = ☐   8 – 8 = ☐   2 – 0 = ☐

What learning will pupils have achieved at the conclusion of Question 4?

- Pupils will have explored subtraction as comparing to find the difference.
- Pupils will have continued to practise using the language of subtraction.

Activities for whole-class instruction

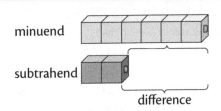

- Remind pupils of the sentence minuend – subtrahend = difference. Explain that there are a number of different ways to subtract. Sometimes we subtract by taking away and sometimes we subtract by comparing two numbers to work out what the difference is.
- Write the calculation 6 – 2. Ask pupils to join six interlocking cubes together in a stick and place the stick on the table in front of them. Then ask them to take two more cubes and join them together in a new stick. They then need to line up the two sticks, matching their left-hand ends together.
- Ask pupils to look along the sticks to see that the first two cubes match; they are in both sticks. They then need to look along the rest of the 6 stick to see that there are four cubes where there are none in the 2 stick. The two sticks are not the same; there is a difference. The difference is the cubes in one stick that have no match in the other stick. Explain that they can think of this as:

 How many more are needed to make the sticks the same.

- Write the matching number sentence: 6 – 2 = 4. Revisit minuend – subtrahend = difference. Pupils found the difference of 4 by comparing the sticks of 6 and the 2.

 4 more were needed to make the sticks the same, the difference is 4.

- Ask pupils to put the 2 stick to one side. Ask them to pick up the 6 stick and remove two cubes from it. The number sentence to record what they did is also 6 – 2 = 4. Subtraction as taking away and subtraction as comparing to find the difference are just two different ways of doing the same thing.
- Pupils should find the differences in Question 4 using the comparing method, making bar models with cubes and drawing them.

Chapter 2 Addition and subtraction within 10

Unit 2.7 Practice Book 1A, pages 55–56

Same-day intervention

- Write the calculation 6 – 2. Ask pupils to join six interlocking cubes together in a stick and place the stick on the table in front of them.
- Remind them that they have explored subtraction (–) as taking away, so they could break two cubes off the stick of 6 to see that 4 are left: 6 – 2 = 4.
- Another way of looking at 6 – 2 is as the difference between 6 and 2. They can think of this as how many more are needed to make the sticks the same.

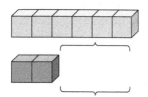

- Ask pupils to make another stick of 2 and line it up below the stick of 6, with both left-hand ends together. Ask how many cubes they need to put with the two cubes to make both sticks the same. Encourage pupils to take some cubes in a different colour and use them to continue the 2 stick until it is the same as the 6 stick. Pupils needed four more cubes, so 6 – 2 = 4, the difference between 6 and 2 is 4.
- Repeat with a similar calculation such as 7 – 6 or 5 – 4. Finally, explore 4 – 4 in the same way. Record the subtraction number sentence each time, using the first number sentence as a model.

Same-day enrichment

- Show pupils a pyramid make from six bricks and explain that they can only use the numbers 1, 2, 3, 4, 5 and 6 within the pyramid. Explain that each number in the pyramid is the difference between the two numbers below it.
- Work through a solution together, for example:

```
      3
    2   5
  4   6   1
```

Check that pupils recognise that 2 is the difference between 4 and 6 because 6 – 4 = 2.
- Tell pupils that there are eight solutions altogether and challenge them to find as many of the other seven solutions as they can. Pupils could work together as a group.

Challenge and extension question

Question 5

> 5 Think carefully and then fill in the boxes. Remember to look for a pattern to help you.
>
> 2 – 0 = ☐ = 3 – 1 = ☐ =
> ☐ – ☐ = ☐ = ☐ – ☐ = ☐ =
> ☐ – ☐ = ☐ = ☐ – ☐ = ☐ =
> ☐ – ☐ = ☐ = ☐ – ☐ = ☐ =

Pupils are asked to find many subtraction calculations with a difference of 2. One example is given.

If they need help, suggest that pupils look at the way the minuend and subtrahend change from the first pair of numbers to the second pair of numbers – can they extend this pattern?

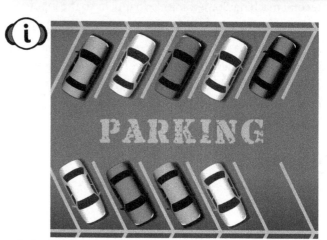

Subtraction does not always require a part to be removed from the whole (taking away). It is also the appropriate calculation when one part of the whole is different from the other in some way and they must be compared – 'finding the difference' requires comparison. In Question 2 above, the example given is that there were there were nine cars in the car park. Four of them were white. So 9 is the whole, the minuend, and 4 is a part, the subtrahend. There is no need for the cars to leave the car park or for anything to be removed from a model. Instead, we need to compare the number of white cars with the total number of cars in order to know how many other cars there are. We can find the value of the difference by counting up from the known part or by subtracting it from the whole, but pupils need to recognise this for themselves through experience with concrete models.

Chapter 2 Addition and subtraction within 10

Unit 2.8
Subtraction (3)

Conceptual context

This is the third in a series of four units focusing on subtraction. In essence, this is a summary unit, revising and consolidating strategies for subtraction since the next unit explores subtraction word problems. The next unit also begins to combine addition with subtraction, with the following three units in this chapter focusing on the inverse relationship between addition and subtraction.

The language of both addition and subtraction is reinforced and pupils continue to explore the link between addition and subtraction. The focus of the unit is on consolidating strategies to solve subtraction number sentences generated in response to pictures, numbers and words.

It is important that, through these activities and questions, pupils have the opportunities to learn that:

- subtractions can be solved using a variety of approaches
- addition and subtraction are closely linked.

Learning pupils will have achieved at the end of the unit

- Pupils will have generated a subtraction number sentence from a picture (Q1)
- Pupils will have consolidated a range of approaches to solve subtraction number sentences, including using addition number bonds, subtraction as taking away and as comparing to find the difference (Q1, Q2, Q3)
- Pupils will have continued to explore the link between addition and subtraction (Q1, Q2, Q3)
- Pupils will have recognised that subtracting zero leaves the minuend unchanged and subtracting the minuend itself gives a result of zero (Q1)
- Pupils will have continued to apply the language of addition and subtraction (Q1)
- Pupils will have consolidated using a variety of representations for recording subtraction, including pictures (Q1), bar models (Q1, Q3), mapping format (Q2), number sentences (Q1, Q2, Q3)

Resources

counting objects, including flowers and two vases, jewels, cubes and counters; mini whiteboards; 12 cm squared cards; sets of 0–9 digit cards; dice

Vocabulary

part, whole, addition, addend, sum, subtraction, take away, minuend, subtrahend, difference

Chapter 2 Addition and subtraction within 10   Unit 2.8   Practice Book 1A, pages 57–58

Question 1

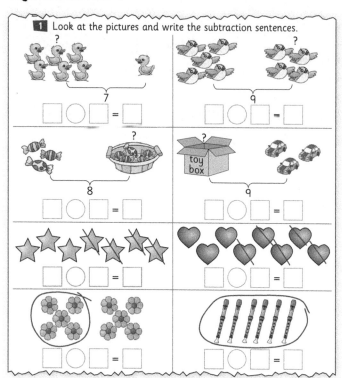

What learning will pupils have achieved at the conclusion of Question 1?

- Pupils will have generated a subtraction number sentence from a picture.
- Pupils will have used concrete and pictorial bar models to reinforce understanding of subtraction using addition bonds.
- Pupils will have developed a bank of strategies to solve subtraction number sentences.
- Pupils have continued to explore the link between addition and subtraction.
- Pupils will have recognised that subtracting zero leaves the minuend unchanged and subtracting the minuend itself results in zero.
- Pupils will have continued to apply the language of addition and subtraction.

Activities for whole-class instruction

- Show pupils two vases containing some flowers. Count the flowers in each vase together, five flowers in the first vase and three in the second. Remind pupils that they can see two parts – that one part is five flowers and the other is three flowers. The whole is eight flowers. Explain that a bar model illustrates this very well.

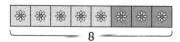

- Ask pupils to write the addition number sentence for the flowers: 5 + 3 = 8. Pointing to the relevant sections of the bar model, ask pupils to say: *5 and 3 are the parts, 8 is the whole.*

- Tell pupils that you have tricked them! What they can see is the situation after subtraction. The two vases are still the parts in the part–whole relationship but originally the flowers were all in one vase. Put the three flowers in with the five, then remove them and ask pupils to write the matching subtraction number sentence: 8 – 3 = 5. Point to the bar model again and ask pupils to say: *5 and 3 are the parts, 8 is the whole.*

- Display the two number sentences together and tell both stories: 5 flowers and 3 flowers, 8 flowers altogether. 8 flowers, 3 put into another vase, 5 flowers left. The two stories are different ways of looking at the same two vases of flowers and the same bar model represents both number sentences.

- Draw a picture of six apples, two on one plate and four on another. Use a bracket to link the two illustrations and label the bracket 6. Draw a matching bar model diagram. Ask pupils to help you complete it with the whole and the parts.

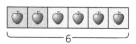

- Ask pupils to write the matching addition and subtraction sentences. Check that pupils have recorded 2 + 4 = 6 or 4 + 2 = 6 and 6 – 4 = 2 or 6 – 2 = 4.

- Draw a new picture of three carrots and a bag of carrots labelled with a question mark next to it. Use a bracket to link the two illustrations and label the bracket 9. Ask pupils to tell you what the diagram is telling them. Confirm that it tells you that the whole is 9 and one of the parts is 3. Draw the matching bar model and record the number sentence as 9 – 3 = ☐.

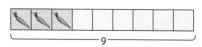

- Ask pupils how they could work out the unknown part. Agree that they may be beginning to remember some number facts – they might recall that 3 and 6 equals 9, so the missing number must be 6. If they solve the

Chapter 2 Addition and subtraction within 10

Unit 2.8 Practice Book 1A, pages 57–58

subtraction number sentence this way, they are using what they know about addition to solve subtraction. As they found out earlier, addition and subtraction are closely linked so they can use one to help them solve the other. Pupils should make a tower of nine cubes, three of one colour and the rest of a different colour. They should talk to each other about how this helps them work out the value of '?'

- Alternatively, they could solve the subtraction number sentence by taking away or by finding the difference, thinking about how many more are needed to make the part the same as the whole.
- Check that pupils recognise that no matter how they solve the number sentence, the missing number is the same.
- Look at Question 1 together. Explain that sometimes the question is about starting with a whole and part is taken away and sometimes they need to show two parts that are separate but are parts of the same thing. Sometimes they can see all the objects but sometimes they will have to work out how many are in the second part by thinking *How many more are needed?* Look at the questions together and identify the different types of question. Ask pupils to complete Question 1.
- After pupils have completed Question 1, ask how they interpreted the recorder question. At first glance it might look like nothing is taken away, so the subtraction number sentence would be 6 − 0 = 6. Looking more closely, all the recorders are taken away, so the picture illustrates 6 − 6 = 0. Challenge pupils to draw an illustration for 6 − 0 = 6.

Same-day intervention

- Draw two simple treasure boxes on a piece of paper. Link them together with a curly bracket as in Question 1. Ideally, use brightly coloured plastic jewels. If these are not available, use cubes or something else to represent jewels
- Give pupils seven jewels to place in the two treasure chests. Remind pupils that they started off with 7 and record 7. When they have finished, say what you see as you record the matching subtraction sentence, for example seven jewels, three in the first treasure chest, which leaves four for the second treasure chest: 7 − 3 = 4.
- Remove the jewels from the treasure chests and ask pupils to put them in the two chests in a different way. Say what they did together and ask pupils to record the matching subtraction number sentence.
- Repeat, changing the starting number of jewels.

Same-day enrichment

- Give pupils six different addition number sentences with a sum of 10 or less. Challenge pupils to find the matching subtraction number sentences.
- Provide a range of counting objects for pupils to choose from to model each addition and help them to find the matching subtractions.
- Ask pupils to choose one of their subtraction number sentences to illustrate.

Question 2

2 Write the correct number in each box.

$8 \xrightarrow{-4} \square$ $10 \xrightarrow{-5} \square$ $6 \xrightarrow{-3} \square$

$5 \xrightarrow{-2} \square$ $7 \xrightarrow{-3} \square$ $8 \xrightarrow{-2} \square$

$\square \xrightarrow{-2} 5$ $\square \xrightarrow{-3} 0$ $\square \xrightarrow{-1} 9$

What learning will pupils have achieved at the conclusion of Question 2?

- Pupils will have consolidated their use of a mapping format for recording subtraction.
- Pupils will have revisited the idea that addition 'undoes' subtraction and that subtraction 'undoes' addition.

Activities for whole-class instruction

- Ask pupils to look at Question 2 and remind them of the function machine they used to take away in Unit 2.7. The recording shown here is just a slightly different way of recording what the machine did.
- Look at the first question: $8 \xrightarrow{-4} \square$.

  Ask pupils to get eight counters and slide them across the table as if they were going through the machine, removing four as they do so. Confirm that they have four counters left: 8 − 4 = 4. Remind pupils that the counters can go back through the machine and the counters removed will be returned, 'undoing' what the machine did the first time. Record the matching addition number sentence: 4 + 4 = 8.

- Show pupils the question: $\square \xrightarrow{-2} 5$

  Check that pupils recognise that the question is showing them the results of some counters going through the machine. Five counters were left. Ask pupils to get five counters, placing them on the table as if they had

103

Chapter 2 Addition and subtraction within 10

Unit 2.8 Practice Book 1A, pages 57–58

been through the machine. Ask what will happen to the counters if they go back through the machine. If necessary, point out that the arrow is labelled – 2. To undo this, they must put the two counters back, in other words add 2. Ask pupils to do this and tell you how many went into the machine initially, 7.

- Ask pupils to complete Question 2, using counters and their imaginary function machine, if needed.

- Ask pupils to revisit Question 2, rewriting each arrow label in two parts. The output will still be the same, they will just be showing it as happening in two stages.

Question 3

3 Complete the subtraction calculations.

8 – 5 = ☐   10 – 4 = ☐   4 – 4 = ☐   6 – 0 = ☐
6 – 3 = ☐   8 – 7 = ☐   9 – 6 = ☐   8 – 3 = ☐
8 – 1 = ☐   5 – 2 = ☐   7 – 6 = ☐   10 – 7 = ☐

Same-day intervention

- Give each pair of pupils eight counters. Ask them to check that they have eight and place them on the table in a 2 by 4 array.
- Each pair will also need a mini whiteboard and a piece of card about 12 cm squared.
- Explain that one pupil is going to operate the machine while the other one records what happened. There are eight counters on the table, so the recorder should write 8. Instead of removing some counters, the machine is going to cover some up. Ask the pupil with the machine to cover up two counters with the card. Ask pupils what the recorder should record. Confirm that the machine covered up two counters, hiding two of them, so the recording should be ──– 2──→. The arrow shows that something happened to the original set of counters and the – 2 shows what it was that happened, two were covered up. Ask pupils to tell you how many counters are left. Confirm that there are six and complete the recording, 8 ──– 2──→ 6.
- Ask pupils to explore and record what happens when the machine covers a different number of the eight counters, regularly swapping roles.
- When pupils are confident with this, change the starting number of counters.

What learning will pupils have achieved at the conclusion of Question 3?

- Pupils will have consolidated a range of approaches to solve subtraction number sentences, including using addition number bonds, subtraction as taking away and as comparing to find the difference.

Activities for whole-class instruction

- Show pupils the outline number sentence ☐ – ◊ = △. Check that pupils recognise that the number sentence has three different-shaped boxes, suggesting that the minuend, subtrahend and difference are three different numbers.
- Give pupils a set of digit cards (0–9) and some counters to use to check their subtractions. Explain that they should use any three digit cards to create a subtraction number sentence that makes sense. There will be some subtractions that they cannot complete, because they only have one card with each number on. For example, if they chose 8 – 4 = △, they would not be able to complete their number sentence.
- Pupils can use addition number bonds, subtraction as taking away or subtraction as comparing to find the difference to help them complete their subtraction number statements, but they should only record those they can make with the digit cards.
- How many different number sentences can the class make?

Same-day enrichment

- Remind pupils that two function machines can be joined together, so that the first one does one thing to the input and the second one does something else. Explain that other machines can do two steps in the same machine.
- Look at the first part of Question 2, 8 ──– 4──→ ☐

and confirm that the solution is 8 ──– 4──→ 4.

Explain that the machine did two different things to the 8 and, together, they resulted in – 4. There are lots of ways that this might have happened, for example ──– 3, – 1──→ or ──+ 6, – 2──→.

Chapter 2 Addition and subtraction within 10

Unit 2.8 Practice Book 1A, pages 57–58

Same-day intervention

- Remind pupils that all the addition and subtraction number sentences they have explored are part–whole relationships.
- Draw a large bar model and complete it for the first calculation in Question 3. Ask: *What is the same and what is different when we compare this bar model to the ones we used before?*
- Write 8 – 5 = ☐ on the board. Explain that the number sentence 8 – 5 = ☐ tells you that the whole is 8 and one of the parts is 5. Remind pupils that the whole is the same size as the two parts.

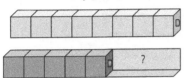

- Ask pupils to make a stick of eight cubes of one colour (for example blue) and a stick of five cubes of another colour (for example orange) and to put them on the table one below the other.
- Ask how they could make the two rows equal. If necessary, add a single (not orange) cube to the orange stick and ask pupils if both rows are the same now. Count the cubes to check. Add another cube and ask the same question, again counting to check. Repeat until both rows contain the same number of cubes. How many were added?
- Return to the number sentence 8 – 5 = ☐. Explain that by finding how many more to make the two rows the same, they found the difference between the two bars: 3. So, 8 – 5 = 3.
- Repeat with a different whole and part to find the other part, recording the subtraction number sentence before and after solving it.
- Discuss how the sticks of cubes look like the bar model – this is why bar models are really good for showing addition and subtraction and helping us to see the missing parts.
- Ask pupils to complete Question 3 in the Practice Book, using sticks of cubes or drawing bar models to help them see the whole and both parts in each question.

Same-day enrichment

- Show pupils a numbered 3 by 3 grid.

1	2	3
4	5	6
7	8	9

- Give pupils counters in two different colours and a dice. Pupils play in pairs, taking turns to roll the dice to identify a difference. They then put a counter on any two numbers, a minuend and a subtrahend that have the same difference as the number shown on the dice. For example, if the dice shows 6, the pupil could put counters on 9 and 3 or 7 and 1 or any other numbers with a difference of 3.
- Each number can only have a single counter on it. When only one number is left without a counter, the winner is the pupil with the most counters on the board.

Challenge and extension question

Question 4

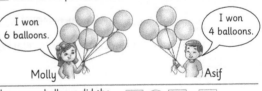

4 Look at the picture and then write the number sentences.

I won 6 balloons.   Molly   I won 4 balloons.   Asif

How many balloons did the children win altogether?	☐ ○ ☐ = ☐
How many more balloons than Asif did Molly win?	☐ ○ ☐ = ☐
How many more balloons does Asif need to win so that he has the same number as Molly?	☐ ○ ☐ = ☐

This challenge and extension question uses one illustration to ask a series of questions that result in addition or subtraction number sentences. If necessary, support pupils to read the questions or read them to pupils. What is important is the way pupils interpret and respond to the questions, not whether they can read them. Pupils should respond to the 'How many altogether?' question with an addition number sentence. The following two questions are essentially 'Find the difference' questions but pupils may not realise this. Encourage them to express the question in a number sentence, considering what they know about parts and the whole. Some pupils may recognise that the second and third questions are asking the same thing in different ways, so the same number sentence will answer both questions.

Chapter 2 Addition and subtraction within 10

Unit 2.9
Let's talk and calculate (II)

Conceptual context

This is the fourth in a series of four units focusing on subtraction. This is a second summary unit, revising and consolidating strategies for subtraction, through word problems and by revisiting previous representations. A further representation for the part–whole relationships, the bar-model, is revisited and practised. This is an excellent image for promoting secure understanding of the part–whole relationship of subtraction. The language of subtraction is reinforced through use of the bar model.

Pupils are exposed to both subtraction and addition in Question 3, effectively revising addition in preparation for exploring the inverse relationship between addition and subtraction in Unit 2.10.

Learning pupils will have achieved at the end of the unit

- Pupils will have practised generating a subtraction number sentence from a story (Q1)
- Pupils will have revisited recording part–whole relationships in abstract representations such as bar models (Q1)
- Pupils will have used concrete and pictorial bar models to reinforce understanding of subtraction using addition bonds (Q1, Q2, Q3)
- Pupils will have consolidated using a variety of representations for recording subtraction, including pictures (Q1), bar models (Q1, Q2, Q3), mapping format (Q2) and number sentences (Q1, Q2, Q3)
- Pupils will have linked the mapping format with the bar model (Q2)
- Pupils will have revisited and reinforced the correct language for subtraction (Q2, Q3)
- Pupils will have revisited recording addition and subtraction in a number sentence (Q3)

Resources

counting objects including small objects, cubes and counters; two cuddly toys; two shallow bowls; dice; squared paper

Vocabulary

part, whole, bar model, addition, addend, sum, subtraction, take away, minuend, subtrahend, difference, table, input, output

Chapter 2 Addition and subtraction within 10

Unit 2.9 Practice Book 1A, pages 59–60

Question 1

> **1** Read each number story and write the subtraction sentence to match.
>
> There were 6 🐦 in the tree.
> 2 🐦 flew away.
> How many 🐦 are left in the tree?
> ☐ ◯ ☐ = ☐
>
> There were 10 🚗 in the car park.
> 7 🚗 drove away.
> How many 🚗 are left now?
> ☐ ◯ ☐ = ☐
>
> There are 8 ✏ and ✒ altogether.
> 5 of them are ✏.
> How many ✒ are there?
> ☐ ◯ ☐ = ☐
>
> There are 7 🐛 and 🐌 in total.
> There are 4 🐌.
> How many 🐛 are there?
> ☐ ◯ ☐ = ☐
>
> There are 9 swans in the lake.
> Four of them are 🦢.
> How many 🦢 are there in the lake?
> ☐ ◯ ☐ = ☐
>
> There are 7 pairs of 🧤 and 🧤 in total.
> Five of them are 🧤.
> How many pairs of 🧤 are there?
> ☐ ◯ ☐ = ☐

What learning will pupils have achieved at the conclusion of Question 1?

- Pupils will have practised generating a subtraction number sentence from a story.
- Pupils will have revisited recording part–whole relationships in abstract representations including bar models.
- Pupils will have used concrete and pictorial bar models to reinforce understanding of subtraction using addition bonds.

Activities for whole-class instruction

- Remind pupils that they can record part–whole relationships in different ways. The bar model gives a very clear picture of what is known and what is unknown; this is the representation they will be using in this session.
- Draw a bar to show the whole, then two smaller bars (that together are the same size as the whole) to represent the parts.

whole	
part	part

- Explain that you are going to tell some stories about swazools to remind pupils how they can use the bar model to help them. Ask pupils if they know what a swazool is. Confirm that you don't know what one is either, and it does not matter because the bar model works for everything.
- Make up a story, for example: *There were seven swazools swimming in the river. Three went home. How many swazools are left swimming in the river?*

7 swimming in river	
3 went home	? stayed

- Ask pupils how they could represent the swazools using connecting cubes. Say: *Show me seven swazools. Now show me three of them going home. Show me the swazools that are left swimming in the river. How many are there?*
- Ask pupils to help you fill in the bar model diagram with the 'knowns' and 'unknowns'.
- Ask pupils if they would complete the diagram differently if the story was about birds, cars or something else. Confirm that it does not matter what the story is about, what matters is what we know about the whole and parts.
- Discuss how to find the missing part. Pupils might recall that 3 + 4 = 7 and use that number fact or subtract 3 from 7 in any way they choose.
- Tell pupils another story about swazools, for example: *In a group of six swazools, four were green. How many were not green?* Ask pupils to draw the matching bar model, for example:

6 swazools	
4 green	? not green

- Invite pupils to place six counters in the whole bar, then move four of them into the part bar labelled 4. Ask pupils how many counters (swazools) they have left to move to the other part bar. Confirm there are 2, so two swazools were not green, 6 − 4 = 2.
- Ask pupils to work through Question 1, drawing bar models and using counters if necessary to help them answer each story and record the matching subtraction number sentence.

Chapter 2 Addition and subtraction within 10

Unit 2.9 Practice Book 1A, pages 59–60

Same-day intervention

- Draw a bar to show the whole, then two smaller bars together, the same size as the whole to show the parts.

whole	
	part

- Tell pupils that you are going to tell them a story about birds and they are going to use cubes to represent birds. Begin the story with, for example: *There were six birds in the tree, this is the whole number of birds.* Ask pupils how they can model this using cubes and the set of bars. Support pupils to arrange six cubes on the set of bars.

part		part		

- Explain that part of the six birds fly away, two birds fly away. Ask pupils how they can show this on the set of bars. Support pupils to move two of the cubes into one of the sections of the bars labelled part. Verbalise this as six birds subtract two birds and record it as 6 – 2.

	6 birds in tree			
2 flew away	? birds left in tree			

- Remind pupils that part of the whole number of birds have flown away, 2 of the 6 birds have flown away, leaving some birds in the tree. Ask pupils how many birds are left – these are the other part of the whole.

	6			
2	4			

- Check that pupils recognise that four birds are left in the tree. Since these are part of the whole, they need to be moved into the other part bar. Verbalise this as six birds subtract two birds equals four birds and record it as 6 – 2 = 4.

- Show pupils the first part of Question 1 in the Practice Book. Explain that this is the story they have just worked through using the bar model. Repeat, modelling 6 as the whole, 2 as a part and 4 as a part as you read the three parts of the question together.

- Work through another of the problems with pupils, modelling it with cubes and bars before asking pupils to work independently.

Same-day enrichment

- Give pupils a partly completed set of bars, for example:

9	
3	part

- Ask them to complete the missing part and use this to write a number story and sentence in the style of Question 1 in the Practice Book.
- Challenge them to go on to complete their own set of bars and matching story.

Question 2

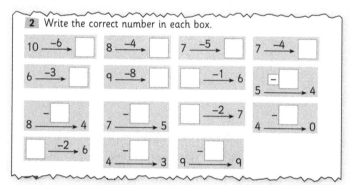

What learning will pupils have achieved at the conclusion of Question 2?

- Pupils will have further consolidated using a mapping format for recording subtraction.
- Pupils will have linked the mapping format with the bar model.
- Pupils will have revisited and reinforced the correct language for subtraction.
- Pupils will have used concrete and pictorial bar models to reinforce understanding of subtraction.

Activities for whole-class instruction

- Remind pupils of the function machine they have used for both addition and subtraction. Draw a table showing the following inputs and outputs:

Input	Output
8	4
5	4
7	4

Chapter 2 Addition and subtraction within 10

Unit 2.9   Practice Book 1A, pages 59–60

- Ask pupils what they notice. Confirm that the output is always 4, whatever is put into the machine. Ask pupils to discuss with their partner what might be happening inside the machine. Share ideas. Confirm that the machine is subtracting a different number each time, so that the output is always 4. Confirm that the input and output are known, but how much the machine is subtracting is unknown.
- Ask pupils to copy the Input/Output table above and add another column to record what the machine is doing each time. Give pupils counters, cubes or small counting objects to use for support.

Same-day intervention

- Remind pupils of the function machine. Draw an example of one, labelling the front with − for subtraction. Explain that the number of objects going into the machine is the whole.

	10 minuend	
part (subtrahend)		part (difference)

- Draw a set of bars for a part–whole model, including the language of subtraction.
- Explain that you are going to put ten objects into the machine. Push ten counters into the machine and amend the bar diagram.

whole (minuend)	
part (subtrahend)	part (difference)

	10 minuend	
6 (subtrahend)		4 (difference)

- Tell pupils that the machine subtracts 6. Record this in the part (subtrahend) bar. Leave six counters in the machine and push the rest out, asking pupils to tell you how many came out of the machine. Complete the difference part of the bar model.

  $10 \xrightarrow{-6} \square$

- Remind pupils that they have recorded what happened in the subtraction function machine in the mapping format before. Record the format next to the bar model and ask pupils to tell you what is the same and what is different. Check that pupils recognise 10 as the minuend and 6 as the subtrahend and that the missing number is 4. Reinforce by placing ten counters in the whole (minuend) section of the bar, then moving six into the subtrahend part and four into the difference part.

All say… *10 is the whole, the minuend; 6 is a part, the subtrahend; and 4 is a part, the difference.*

- Draw some mapping format subtractions with missing difference and ask pupils to draw the matching bar model and use it to find the missing number. Provide counters to use with the bars. Show pupils an arrow format subtraction with the subtrahend missing, for example:

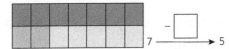

Compare with the previous arrow format, exploring what is the same and what is different. Confirm that the subtrahend is missing, but the difference is known. Draw the matching bar model so that pupils are clear about what they know and what they don't know.

	10 minuend	
? (subtrahend)		5 (difference)

- Ask pupils to find the unknown subtrahend, using the bar model for support.
- Draw some arrow format subtractions with missing subtrahends and ask pupils to draw the matching bar model and use it to find the missing number. Provide counters to use with the bars.

Same-day enrichment

- Give pupils a partly completed table headed with Input (minuend), Output (difference) and Subtraction (subtrahend).

Input (minuend)	Output (difference)	Subtraction (subtrahend)
8		7
	2	5
5		5
9	7	
6		3

- Challenge them to complete the table, using any method they choose to help them. Ask pupils to draw a matching bar model for all the rows of the table.

Chapter 2 Addition and subtraction within 10

Unit 2.9   Practice Book 1A, pages 59–60

Question 3

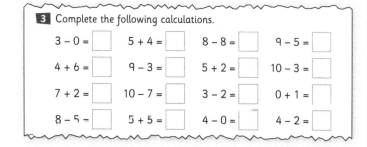

What learning will pupils have achieved at the conclusion of Question 3?

- Pupils will have revisited recording addition and subtraction number sentences.
- Pupils will have further reinforced the correct language for subtraction.
- Pupils will have used concrete and pictorial bar models to reinforce understanding of subtraction using addition bonds.

Activities for whole-class instruction

- Tell pupils that they have now investigated addition and subtraction and have explored different ways for finding any missing number in an addition or subtraction number sentence. Remind them of the format of the number sentence:

- The square boxes always contain numbers and the circular box always contains a symbol, + for addition and – for subtraction.
- Tell pupils they are going to play a game. In pairs, they should roll a dice twice and write the two numbers in the first two square boxes. If the first number rolled is the larger number, they should subtract the second number. If the first number is the smaller number, they should add the two numbers. Whether they add or subtract, they must record the matching number sentence.
- Encourage pupils to draw a bar model to help identify the parts and whole for any of their calculations. Provide pupils with some small counting objects, counters or cubes and squared paper to use to support their thinking and calculations.

Same-day intervention

- Show pupils two cuddly toys, two shallow pots, some counters and a dice. Give each cuddly toy a shallow pot and place the counters on the table between them. Ask pupils to take turns to roll the dice for each toy.
- When the dice is rolled, pupils agree the amount shown on the dice and give the toy that many counters. Record the number for each toy as the first addend in a number sentence.
- Give each toy a second turn, adding the counters to their pot. Record the number as the second addend, for example 3 + 6 = ☐. Remind pupils that + means add and ask them to explain what that is.
- Help each toy to find the sum of their counters by removing them from the pot and laying them out next to the number sentence. Encourage pupils to read the first number and count on the rest of the counters to find the total.
- Once pupils have identified each toy's number of counters, record the total. Begin a subtraction number sentence with the sum as the minuend.
- Explain that each toy is going to roll the dice again, but they will subtract the amount shown on the dice this time. Wonder aloud which toy will have the most counters left. Will it be the one who has the most now? Which one is that? Or will it be the one who has fewer counters now? Which one is that?
- Roll the dice for the first toy and record the subtraction number sentence, for example 9 – 2 = ☐. Remind pupils that – means subtract and ask them to explain what that is.
- Invite pupils to tip out and count the counters and then take away the correct number to find out how many are left. Some pupils may be able to count to 9 and count back 2 as they remove two counters one at a time. Complete the number sentence. Repeat with the second toy. Compare the number of counters each toy has left.
- Repeat the activity with pupils taking the lead.

Chapter 2 Addition and subtraction within 10

Unit 2.9 Practice Book 1A, pages 59–60

Same-day enrichment

- Give pupils a dice and ask them to roll the dice twice. If the first number rolled is the larger number, they should subtract the second number. If the first number is the smaller number, they should add the two numbers. Whether they add or subtract, they must record the matching number sentence.
- Pupils then use the total as the starting number in their next number sentence. They only need to roll the dice once more to create a new number sentence. Challenge them to make their list of number sentences as long as possible, for example:

  9 − 2 = 7
  7 − 4 = 3
  3 − 1 = 2
  2 + 2 = 4
  4 + 5 = 9 …

Challenge and extension question

Question 4

4 What does each shape stand for?

If △ + ☆ = 10 and 8 − ☆ = 2, then
△ = ☐ and ☆ = ☐.

If ◆ − ● = 7 and ● + 3 = 5, then
◆ = ☐ and ● = ☐.

This question requires pupils to use what they know about addition and subtraction sentences and the relationship between them to work out what a symbol within a sentence represents. They can then substitute the value of that symbol into another number sentence to work out the value of a different symbol.

Pupils need to recognise that, where there is only one unknown in one of the number sentences, they can work out the missing number. For the first pair of number sentences, they can use 8 − ☆ = 2 to identify the value of the star. Pupils may use anything they like to help them, though they may also be able to work mentally. Having identified the value of the star, pupils need to substitute its value into the other number sentence to find the value of the triangle. The process is the same in the second pair of number sentence, but whereas the first set began with an addition, the second set begins with a subtraction.

Chapter 2 Addition and subtraction within 10

Unit 2.10
Addition and subtraction

Conceptual context

This unit focuses on the inverse relationship between addition and subtraction. Pupils have encountered this relationship previously where subtraction was seen to 'undo' or reverse addition and addition was seen to 'undo' or reverse subtraction. In this unit, these ideas are developed. The inverse relationship is represented through a variety of models and images including the bar model, mapping diagrams and other representations of part–whole relationships. Understanding of the inverse relationship is then applied to generate four related facts, two additions and two subtractions. The language of addition and subtraction is reinforced throughout.

All questions in the unit focus on addition and subtraction, sometimes in the same question (Questions 2 and 4).

Learning pupils will have achieved at the end of the unit

- Pupils will have modelled part–whole relationships with bar models and drawings (Q1, Q2, Q3)
- Pupils will have generated addition and subtraction number sentences from the same part–whole representation (Q1, Q2, Q3)
- Pupils will have revisited and linked the mapping representation with the bar model (Q2)
- Pupils have revisited the link between addition and subtraction (Q2)
- Pupils will have revisited and reinforced the correct language for addition and subtraction (Q3, Q4)
- Pupils will have explored two-step calculations using addition and/or subtraction in any order (Q4)
- Pupils will have extended using a mapping format to record addition, subtraction and mixed calculations (Q4)
- Pupils will have revisited and consolidated the idea that addition 'undoes' subtraction and reminded that subtraction 'undoes' addition (Q4)

Resources

books; linking toys; interlocking cubes; sticky notes; card rectangles (5 cm by 15 cm); card squares (4 cm by 4 cm) with triangle arrow; construction bricks; two shoeboxes

Vocabulary

part, whole, bar model, addition, addend, sum, subtraction, take away, minuend, subtrahend, difference, table, input, output, row, doing, undoing

Chapter 2 Addition and subtraction within 10

Unit 2.10 Practice Book 1A, pages 61–63

Question 1

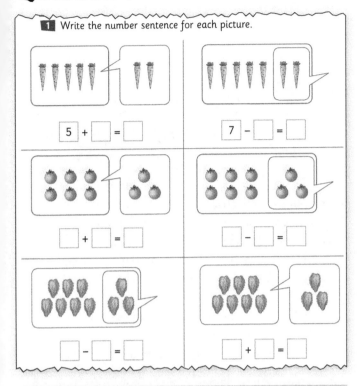

What learning will pupils have achieved at the conclusion of Question 1?

- Pupils will have modelled part–whole relationships with bar models and drawings.
- Pupils will have generated addition and subtraction number sentences from the same part–whole representation.
- Pupils will have revisited the link between addition and subtraction.
- Pupils will have revisited and reinforced the correct language for addition and subtraction.

Activities for whole-class instruction

- Show pupils a stack of five similar books. Ask pupils whether they would like to add some more or take some away. If adding, confirm that the existing stack is a part. Make a new stack of three books, ideally a different colour, confirming that this is the other part. Stand both stacks of books upright and ask pupils if the row of books reminds them of anything. If necessary, suggest that it is like the two parts in a bar diagram but this one is vertical. Turn the pile of books onto its side so that pupils can see the connection between the vertical and the horizontal representations.
- Ask pupils to draw a bar model, labelling the whole and the parts. Ask pupils what the parts and whole are called in addition (addends). Read the bar diagram together: 5 is a part, 3 is a part, 8 is the whole and 5 is an addend, 3 is an addend, 8 is the sum. Confirm that the matching addition sentence is 5 + 3 = 8.

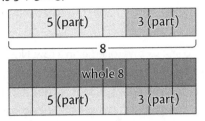

- Either of these bar models is acceptable. The first is a better representation of a simple part–part–whole situation when both parts are present. The second is better when one part is missing.

- Remind pupils that there are lots of other illustrations they could use, for example each part could be drawn in a box and an arrow drawn to join the two boxes.
- Ask pupils to draw their own representations of 5 things add 3 things equals 8 things. Share a few of the different representations.
- Return to the original stack of five books. Stand them up and explain that you are going to subtract some of the books, so what pupils can see at the start is the top row of the bar model, the whole. Turn the pile of books on its side again. Ask pupils to draw this and then complete their bar diagram, deciding for themselves what the two parts will be.

whole 5	
(part)	(part)

- Share ideas for the value of the two parts, separating the five books according to the values given. Ask pupils to draw a matching bar model diagram, labelling the whole and the parts. Ask pupils what the parts and whole are called in subtraction. Choose the value of the two parts and read the bar diagram together, for example: *3 is a part, 2 is a part, 5 is the whole and 5 is the minuend, 3 is the subtrahend and 2 is the difference.* Confirm that the matching subtraction sentence is 5 − 3 = 2.
- Ask pupils to imagine situations using real-world objects in which a whole, made up of a number of those objects, can be partitioned into two parts in different ways. Suggest they set up manipulatives to model situations.
  - Can they represent their parts being put together to create the whole by drawing something? By writing a number sentence?
  - Can they represent their whole being partitioned to create the parts by drawing something? By writing a number sentence?
- Share a few of the different representations.

Chapter 2 Addition and subtraction within 10 Unit 2.10 Practice Book 1A, pages 61–63

Same-day intervention

- You will need some toys that link together such as elephants, links or S-shapes. Alternatively, use interlocking cubes.
- Ask pupils to link six toys together in a row. Now ask pupils to link together five toys, ideally using a different colour. Pupils should line up both sets so that they start from the same point, comparing the length of the two rows.
- Ask pupils what is needed to make both rows the same length. Confirm that one more toy needs to be added to the row of 5. Instead of attaching it, lay it next to the row of 5.
- Show pupils that they have made a bar model. The first row is the whole, 6, the next row is the parts, 5 and 1.

 6 is the whole, 5 is a part, 1 is a part.

- Remind pupils that they can write an addition sentence to show the part + part = whole.
- Ask pupils to arrange the two parts with a small gap between them. Give them a sticky note with + on to place between the two parts. Then give pupils an equals sign = on a sticky note to place after the parts and before the whole. Read the number sentence together: 5 + 1 = 6. Ask pupils to swap the two addends around and read the addition number sentence: 1 + 5 = 6.
- Ask pupils to make the bar model again. This time, ask them to remove the whole and set it out with a sticky note with the subtraction sign on it next. Tell pupils they are going to subtract 1 from the six cubes and ask them to position the correct set of toys after the subtraction sign. Give pupils an equals = sign on a sticky note and ask them to complete their number sentence. Read the sentence together: 6 − 1 = 5.
- Remind pupils that they have made the number sentences from the same arrangement of toys because addition and subtraction are related. Addition undoes subtraction and subtraction undoes addition, so they can both be represented by the same set bar model or the same set of objects.

Same-day enrichment

- Give pupils two cards each, one rectangle 5 cm by 15 cm, and one 4 cm squared with a small triangle on one edge to make it look like the addition and subtraction pictures in Question 1.

- Show pupils how to place the square card on top of the rectangle with the triangle pointing away from the rectangle to show subtraction, and the square placed next to the rectangle but pointing to it to show addition

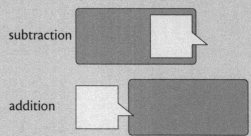

- Give pupils ten beans, counters or cubes and challenge them to create a mixture of additions and subtractions, using the cards and beans to illustrate each calculation and recording each in a number sentence. Explain that you would like the sum or differences for their calculations to use all the numbers from 0 to 10 once and only once. Check that they recognise that this means they must create 11 calculations. Suggest that they may be able to use a similar layout for an addition and subtraction. Explain that it is up to them how many beans they use each time. They do not need to use them all every time.
- If necessary, show pupils Question 1 in the Practice Book to help them get started.

Question 2

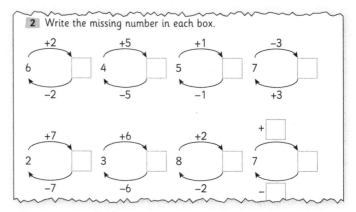

What learning will pupils have achieved at the conclusion of Question 2?

- Pupils will have further developed a mapping representation to link addition and subtraction.
- Pupils will have revisited and linked the mapping representation with the bar model.
- Pupils will have modelled part–whole relationships with bar models and drawings.

Chapter 2 Addition and subtraction within 10

Unit 2.10 Practice Book 1A, pages 61–63

- Pupils will have generated addition and subtraction number sentences from the same part–whole representation.
- Pupils will have revisited the link between addition and subtraction.

Activities for whole-class instruction

- Revisit the first bar model diagrams used in Question 1. Ask pupils to use eight cubes to make a stick representing the whole, five cubes in a different colour to represent one part and three cubes in a different colour to represent the other part, then lay out the sticks of cubes to make their own bar diagram.

	5 (part)		3 (part)	
		8		

- Challenge pupils to record + and = on sticky notes and use these and the cube sticks to lay out the number sentence.
- With the addition number sentence in front of pupils, draw an arrow representation similar to those used in Question 2 alongside the bar model.

- Remind pupils that they already know that addition can be 'undone' by subtraction. Ask pupils to explore using their sticks of cubes to change their addition sentence into a subtraction sentence. Suggest they make a stick of eight cubes to represent the whole and place it above the parts to make a bar model that looks like the one below (they have seen one like this previously). Then, when one is removed, seeing the whole will help them see how big the missing part is.

	whole 8			
	5 (part)		3 (part)	

- Confirm that 8 – 3 = 5 'undoes' the addition. Ask how you could add to the arrow diagram to show this.
- Complete the arrow diagram together.

Same-day intervention

- Make an A4 version of an empty double arrow diagram as used in Question 2 for each pupil.

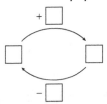

- Ask pupils to make a tower of four cubes of the same colour to place in the left-hand square. Trace the top arrow with a finger, pausing at the + box to say *add*. Ask pupils how many cubes they would like to add to the four cubes. Agree a number such as 2 and ask pupils to make a tower of that many cubes of a different colour to place in the top box.
- Connect the two towers together as you say *add* 2 and then, as you say *equals*, move the connected tower into the right-hand box.
- Read the top arrow calculation together, 4 + 2 = 6, recording it in a number sentence.
- Look at the bottom arrow and tell pupils they are going to 'undo' or reverse the calculation they have just done. Point out the – sign next to the bottom box. Explain that this shows that the tower will be split into two parts in this box. Ask pupils how it should be partitioned.
- They should see the two different colours in the tower of 6 and remember that they added 2 when they completed the addition calculation, so should know to 'subtract 2' or 'take away 2'.
- Point out that the 4 tower is left and place this in the left-hand box. Read the bottom arrow calculation together: 6 – 2 = 4.
- Read around the whole circle two or three times so that pupils recognise that they are literally going round in circles. They are adding and subtracting the same amount so that they are back at the start. Check that pupils recognise that the subtraction is 'undoing' the addition.
- Finally, take the towers of 6, 4 and 2 and use them to make a bar model. Read the addition and subtraction number sentences again, but this time, point to the bar model diagram. Ask pupils to rearrange their cubes and repeat with you.

	whole 6		
	4 (part)	2 (part)	

Chapter 2 Addition and subtraction within 10

Unit 2.10 Practice Book 1A, pages 61–63

- Explain that the arrow format and the bar model are just two representations of part–whole relationships, so we can see and record the same number sentences from both.
- Ask pupils to use their empty double arrow diagram to model and complete Question 2.

Same-day enrichment

- Remind pupils that they already know that addition 'undoes' subtraction and subtraction 'undoes' addition.
- Agree an addition such as 5 + 4. Model it with towers of cubes, then rearrange the towers to form a bar model. Remind pupils of the arrow format used in Question 2 and ask them to record the number sentence and the matching arrow diagram.

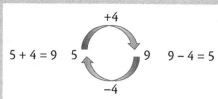

- Use the bar model and towers of cubes to find the 'undoing' subtraction. Ask pupils to use an arrow in the opposite direction to record this, along with the matching number sentence.

- Finally, ask pupils to draw the matching bar model alongside the arrow format and number sentences. Pupils used their towers of cubes to model this earlier so should find this straightforward.
- Challenge pupils to complete Question 2, recording the two number sentences next to each diagram and drawing corresponding bar models. Provide them with blank bar models like these:

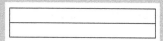

Pupils must decide where to draw the line separating the parts in the bottom section, depending on the relative sizes of the parts. This only needs to be approximate.

Question 3

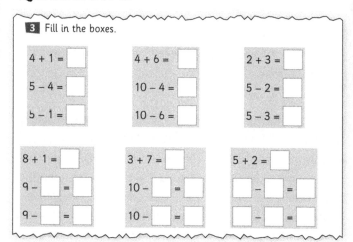

What learning will pupils have achieved at the conclusion of Question 3?

- Pupils will have revisited and reinforced the correct language for addition and subtraction.
- Pupils will have modelled part–whole relationships with bar models and drawings
- Pupils will have generated four addition and subtraction number sentences from the same part–whole representation.
- Knowledge about links with related facts from any number sentence will have strengthened pupils' number concepts and developed their ability to think flexibly.

Activities for whole-class instruction

- Ask pupils to make sticks of nine, five and four cubes using different coloured cubes for each stick. Ask pupils to tell you how to turn the towers into a bar model.
- Make up the model as instructed and read it together, for example: *9 is the whole, 4 and 5 are the parts.*

whole 9	
part 5	part 4

- Ask pupils to put the information they have just said into number sentences. Explain that there are four number sentences for each part–whole relationship and you would like pupils to find all four.
- Give pupils a few minutes and then remind them that for addition, addend + addend = sum. Ask pupils to suggest the two addition number sentences for this bar model. Confirm that these are 4 + 5 = 9 and 5 + 4 = 9 and record them.

Chapter 2 Addition and subtraction within 10

Unit 2.10 Practice Book 1A, pages 61–63

- Remind pupils that, for subtraction, minuend − subtrahend = difference. Ask pupils to suggest the two subtraction number sentences for this bar model. Confirm that these are 9 − 5 = 4 and 9 − 4 = 5. Record these with the addition number sentences.
- Explain that these number sentences are the family of four facts for this bar model. The same part–whole relationship could be represented with arrows, a picture or something else but however it is represented, the same four number sentences can be used to record the part–whole relationship.

Same-day intervention

- You will need some interlocking construction bricks of the same size but different colours. Build a tower with six bricks the same colour and lay it down so that it looks like a bar. Make two more towers in different colours, one with four bricks and one with twp bricks. Lay these down to make the parts row of the bar model.
- Tell pupils that you have made a bar model. Point to the relevant parts as you say: *6 is the whole, 4 is a part and 2 is a part*. Ask pupils to repeat this with you.
- Draw around the bar model, carefully labelling each part with its numerical value.

  (i) The act of physically drawing around the cubes to create the connection between concrete and pictorial representations will help pupils to strengthen concepts and deepen understanding.

- Ask pupils to take turns to carefully draw around your bar model, labelling the whole and the parts.
- Using +, − and = on sticky notes, remove the pieces from the bar model one at a time to create the number sentence layout of 2 bricks + 4 bricks = 6 bricks. Pupils record this next to their bar model as 2 + 4 = 6. Swap the two part bars around to make 4 bricks + 2 bricks = 6 bricks and ask pupils to record this as 4 + 2 = 6.
- Return the bricks to the bar model outline and then rearrange to create the subtraction number sentences. Pupils record the matching subtraction number sentences, 6 − 4 = 2 and 6 − 2 = 4. Name the minuend, subtrahend and difference with pupils, repeating each name.
- Read all four number sentences together, pointing to the relevant part of the bar model diagram. Confirm that the bar model is a useful representation for addition and subtraction because they both express part–whole relationships.

Same-day enrichment

- Give pupils a bar model diagram labelled with two parts and a whole.

whole	
part	part

- Ask them to use the bar model to help them find four calculations (two additions and two subtractions) for each row of numbers in the table.

part	part	whole
2	7	9
5	3	8
6	0	6
1	6	7
8	2	10

Question 4

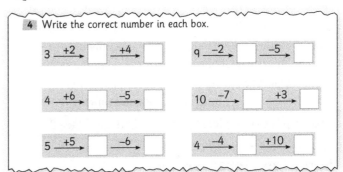

What learning will pupils have achieved at the conclusion of Question 4?

- Pupils will have extended using a mapping representation to record addition, subtraction and mixed calculations.
- Pupils will have explored two-step calculations, using addition and/or subtraction in any order.
- Pupils will have revisited and reinforced the correct language for addition and subtraction.
- Pupils will have revisited and consolidated the idea that addition 'undoes' subtraction and reminded that subtraction 'undoes' addition.

Activities for whole-class instruction

- Prepare two shoeboxes to use as a double function machine, or reuse your previous function machine with a shoebox.

Chapter 2 Addition and subtraction within 10

Unit 2.10 Practice Book 1A, pages 61–63

- Remind pupils that they have used function machines before. Some objects go into the machine, the input, things are added or taken away, and the result, the output, comes out of the machine.
- Explain that two machines can be joined together, so that two things can happen to whatever is put into the machine. Make a stick of six cubes and put it into the first machine. Add one cube and push the stick out. Show pupils and ask them what has happened. Confirm that one cube has been added, so we could record what happened as: $6 \xrightarrow{+1} 7$. Explain that the 7 is the input for the second machine.
- Push the stick of seven cubes into the second machine, remove three and push out four cubes. Ask pupils what has happened and how you could record it. Agree that you can use what you recorded for the first machine and extend it, to show what happened altogether: $6 \xrightarrow{+1} 7 \xrightarrow{-3} 4$.
- Ask pupils to use two pieces of paper to represent the two machines. Working with a partner, one pupil guides an input through the machine in the form of a stick of cubes, carries out an action in each machine in turn and guides the final output out of the second machine. Meanwhile, their partner records what happened using the arrow format. Pupils swap roles and repeat.
- Ask pupils to complete Question 4 in the Practice Book.

Same-day intervention

- Use two shoeboxes as function machines, but with the lids removed so that pupils can see what is happening. Make hand-sized holes in the side walls of both boxes to use as inputs and outputs.
- Show pupils the two machines and ask them to decide what each one does. Clearly label the front of each machine with an arrow to show the direction from input to output and + number or − number.
- Choose an input number and ask a pupil to make a tower of that many cubes and put it through the first input. Ask another pupil to confirm what the machine does and to do that to the tower of cubes before pushing it out through the first output. Record what happened, for example $3 \xrightarrow{+2} 5$, saying: *Three cubes went into the machine; the machine added two cubes; five cubes came out of the machine.* Ask pupils to repeat this with you as you point to each part of the recording.
- Ask another pupil to pass the stick of cubes through the second input and then carry out the second action before pushing the stick of cubes out of the second output. Record what happened, linking to the recording for the first machine, for example $3 \xrightarrow{+2} 5 \xrightarrow{-1} 4$, saying: *three cubes went into the machine. The first machine added two cubes, five cubes came out of the first machine five cubes went into the second machine, the second machine subtracted one cube, four cubes came out of the second machine.* Ask pupils to repeat this with you as you point to each part of the recording.
- Agree changes to what each machine does with pupils and re-label the machines. Repeat the action of the paired machines with a different input, recording and saying what happened as before.

Same-day enrichment

- Challenge pupils to add a third action to each of the parts of Question 4 so that every final output is the same as the initial input.
- Remind pupils that addition 'undoes' subtraction and subtraction 'undoes' addition. It will be helpful to look at what has already happened to the input and consider how that can be reversed in a single further action.
- Encourage pupils to use three pieces of paper to represent the three actions, passing a tower of cubes along to see the effect of the first two actions.

Challenge and extension question
Question 5

5 Write subtraction calculations with a difference of 3.

☐ − ☐ = 3   ☐ − ☐ = 3

☐ − ☐ = 3   ☐ − ☐ = 3

☐ − ☐ = 3   ☐ − ☐ = 3

☐ − ☐ = 3   ☐ − ☐ = 3

Pupils are asked to find subtraction calculations with a difference of 3. There are an infinite number of solutions and the eight solutions recorded may follow a pattern or be random. If pupils find it hard to know where to start, suggest they begin with 3 − 0 = 3 and then consider how they can change this to ensure that the difference of 3 is maintained. Pupils will need to recognise that they must increase the minuend and the subtrahend by the same amount to maintain a difference of 3.

Chapter 2 Addition and subtraction within 10

Unit 2.11
Addition and subtraction using a number line

Conceptual context

This unit focuses on using a number line for addition and subtraction. Pupils were introduced to the number line in Unit 1.14 where they completed the missing numbers and explored jumping along it in jumps of equal size. This unit moves pupils on to calculating.

Pupils are familiar with different representations and structures for addition and subtraction calculations with missing parts or whole. In this unit, pupils will explore how to solve each of these using a number line.

Learning pupils will have achieved at the end of the unit

- Pupils will have constructed their own 0–10 number lines (Q1)
- Number lines will have been introduced as a tool to find the sum or difference in a number sentence (Q2, Q4)
- Pupils will have explored finding a missing number in an addition or subtraction sentence (Q3, Q4)

Resources

sheets of paper; interlocking cubes; counters; counting objects; rulers; wallpaper backing paper; chalk; clipboards; sets of 0–10 number cards; 0–10 number lines

Vocabulary

number line, add, addition, addend, sum, subtract, subtraction, minuend, subtrahend, difference

Chapter 2 Addition and subtraction within 10

Unit 2.11   Practice Book 1A, pages 64–65

Question 1

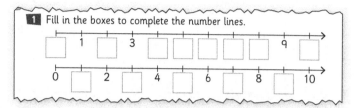
1 Fill in the boxes to complete the number lines.

What learning will pupils have achieved at the conclusion of Question 1?
- Pupils will have revisited and recorded the order of numbers from 0 to 10.
- Pupils will have constructed their own 0–10 number lines.

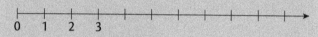

Activities for whole-class instruction

- Give pupils ten interlocking cubes, a sheet of paper and a ruler. Ask pupils to make the ten cubes into a stick and lay it down on the paper, leaving plenty of room around the stick. They then use the ruler to draw a straight line about the same length as the stick of cubes.
- Show pupils how to line up the stick of cubes with the line they have drawn. Ask them to draw a mark on the line where the first cube starts and label it 0, because there are no whole cubes yet. At the place where the first cube joins the second, they should mark the line again and label it 1 as there is one whole cube at this point. Pupils continue marking the line at the end of each cube up to 10, then remove the stick of cubes.
- Tell pupils that they have made their own number line. Each number they have marked shows how many cubes there were at that point on the line. Ask pupils to break three cubes off their stick of ten and line them up with 0. The end of the stick should be at 3.

     The 3 mark shows there were three cubes up to this mark.

    Repeat with other numbers.
- Count along the number line together. What pupils have is a useful line of numbers, all in the right order, that they will be able to use to help them add and subtract.
- Ask pupils to complete Question 1 in the Practice Book.

    Number tracks have a single number in each space, supporting understanding of one-to-one correspondence between numbers and squares. They are useful for developing understanding of ordinal and cardinal number. Number tracks usually start at 1 and extend to 10 or 20. A 100-square is an extension of a number track.

Number lines start from 0. Although they are also used for developing understanding of ordinal and cardinal number, they also show continuity from one number to the next, rather than the separation suggested by a number track. This sense of continuity is important for a deep understanding of how numbers work when calculating and measuring. Numbers are shown on the lines, not in the spaces, on a number line; this shows that the value is complete at that point. A number line is a more accurate representation than a number track.

Young pupils start to learn about numbers using number tracks but begin to use number lines in Year 1.

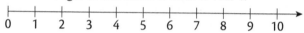

Same-day intervention
- Give 11 pupils a single 0–10 number card each. Ask the other pupils to help arrange pupils in order from 0 to 10 to make a human number line.
- Count along the number line together from 0 to 10 and back to 0. Show pupils an enlarged 0 to 10 number line on the floor and explain that pupils have made a human number line.
- Put the number cards away. Ask a pupil to stand on the 1 mark on the floor number line and jump along the number line, making a single jump from one number to the next. Ask the pupil to stop at 6 and ask pupils how many more jumps are needed to reach 10. Agree that 4 more are needed, because 6 + 4 = 10. Check that pupils understand the marks are where the numbers are shown, not the spaces.
- Ask another pupil to jump from 10 to 7 in single jumps. Ask pupils how many more jumps along the number line to reach 4. Agree that 3 more are needed, because 7 − 3 = 4.
- Give each pupil a copy of a 0–10 number line and some cubes. Ask them to make a tower of cubes to match each number, placing them near the relevant numbers along the number line.
- Count along the towers from 0 to 10 together. Check that pupils recognise that the numbers are growing by 1 as they count from each tower to the next. Count back from 10 to 0 together, checking that pupils recognise that the numbers are reducing by 1 as they count back from each tower to the next.

Chapter 2 Addition and subtraction within 10

Unit 2.11 Practice Book 1A, pages 64–65

Same-day enrichment

- Give pupils some number lines of different lengths without numbers.
- Write just one number on the number line and draw boxes next to other marks for pupils to complete.

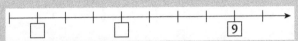

- Vary the position of the given number and the boxes to complete, as well as the length of the line and the number of marks on the number line.
- Pupils may initially assume that the left-hand end is always 0 and the right-hand end is 10. Give them some lines with one of the end points labelled with a number other than 0 or 10 to ensure they check the number line carefully.

Question 2

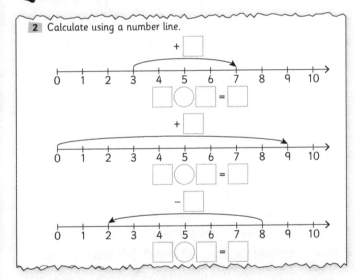

What learning will pupils have achieved at the conclusion of Question 2?

- Pupils will have been introduced to using a number line to support finding the sum in a number sentence.
- Pupils will have been introduced to using a number line to support finding a difference in a number sentence.

Activities for whole-class instruction

- Show pupils a large version of the 0–10 number line. Make sure each pupil has their own number line and some cubes.
- Ask pupils to make a stick of two cubes and to lay it on their number line so that the stick ends at 2. Now ask pupils to make another stick of three cubes in a different colour and join it to the stick of two cubes. Ask pupils where the stick of cubes ends on their number line and how many cubes are in the stick. Both numbers should be 5.
- Ask pupils to move the cubes away from the number line, but to ensure the cubes still align with it. Remind them that they first made a stick of two cubes, so they can put a small mark on the number line at 2. They added three more cubes. Ask them to count on 3 more from 2, ending on 5. Ask pupils to put a small mark on 5 and show them how to draw a curved arrow from 2 to 5 on their number line to show they added three more cubes. Pupils label the jump just above the arrow with + 3.
- Ask pupils to look at both their cubes and their marked number line to write the matching number sentence. They started with two cubes, so 2 is an addend. They added three more, so 3 is the other addend and they need + to show that they added. The sum was 5, so they can record = 5, 2 + 3 = 5. Check that pupils recognise that they could have used the number line without any cubes, just by drawing the arrow to find the sum.
- Tell pupils they are going to use the number line for subtraction. Ask pupils to make a stick of nine cubes. Pupils should line them up against their number line to check.
- Ask pupils to subtract two cubes from their stick and check where the stick comes to on their number line. It should be seven cubes long, the end of the stick lining up with the 7, because 9 – 2 = 7. Ask pupils to draw in the arrow from 9 to 7, writing – 2 above the arrow. They can then record the number sentence: 9 – 2 = 7.
- Ask pupils to look again at the arrows they have drawn. The first arrow showed adding 3. To find out the total, they need to count on 3 from 2, reaching 5: 2 + 3 = 5.
- Ask pupils to look at the arrow from 9 to 7. This shows that they started with 9 and took away 2 to leave them with 7. This is subtraction by taking away.

Same-day intervention

- Make a large floor 0–10 number line. (Wallpaper backing paper is good for this.) Alternatively, chalk a number line outside.
- Count along the number line together, both forwards and back. Ask pupils to take turns to take one jump to the next number all the way along the number line from 0 to 10.
- Choose two pupils to play a game. Both start at 0 and take turns to roll the dice and jump on that number of spaces. Their aim is to reach 10. Have a clipboard with

Chapter 2 Addition and subtraction within 10

Unit 2.11   Practice Book 1A, pages 64–65

two 0–10 number lines on and ask pupils to help you record each turn. Encourage some pupils to take over the recording. Alternatively, pairs could have clipboards to record together.

- If pupils would go beyond 10 in a turn, then they must miss that turn. It would not be possible to record jumps beyond 10 on a 0–10 number line and would only confuse.
- Repeat the game until all pupils have had a turn. Experiencing moving along the line is a useful way to develop understanding of the number line.

Same-day enrichment

- Give pupils some small counting objects and ask them to show you 5 + 3 and then show you the sum. Talk about the different arrangements and check that pupils are confident that 5 + 3 = 8.
- Give pupils a 0–10 number line or ask them to use the one they made. Remind them that they can count to 10 along the number line, just as they can count ten objects. The numbers on the line are in the same order as we say them and we can use that to help us calculate.
- Refer to the calculation that they just did and mark 5 on the number line. Remind them that they added 3 more, so count along 3 more places, drawing the arrow from 5 to 8 as you go. Label the arrow + 3. Point out that you finished on 8: 5 + 3 = 8. The result on the number line is the same as when they used objects.
- Give pupils two addition number sentences, for example 6 + 2 = ☐ and 4 + 3 = ☐, to complete using the number line. Ask them to draw in the arrow and label it so you can see how they solved the sentences. Pupils could rub out the first arrow once they have discussed it with you or use a different coloured pencil for the second arrow.
- Ask pupils how they might use the number line to support subtraction as take away. Work through an example together, such as 9 – 3 = ☐, drawing the arrow, labelling it and confirming that 9 – 6 = 3.
- Ask pupils to complete Question 2 in the Practice Book.

Question 3

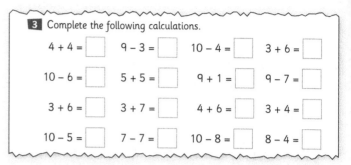

What learning will pupils have achieved at the conclusion of Question 3?

- Pupils will have explored using a number line to find a missing addend.
- Pupils will have explored using a number line to find a missing subtrahend.

Activities for whole-class instruction

- Ask pupils to work in pairs. Give each pair a set of 0–10 number cards and 0–10 number lines (dry-wipe or paper).
- Show pupils an addition layout.

- Explain that they need to separate their cards into two piles. The first pile is 0–5. They need to shuffle these and place them face down over the first number box. The second pile will be 6–10. These should be shuffled and placed over the last number box.
- Pupils turn over the top card in each pile and then use the number line to find out how many more are needed to reach the sum given by turning over a card in the last box.
- Work through an example together: 4 + ☐ = 9. Find 4 on the number line and draw an arched arrow to 9. Count how many to get from 4 to 9 and complete the number sentence, 4 + 5 = 9.

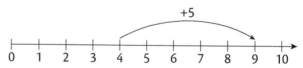

- After pupils have completed five calculations, ask them to shuffle the cards and return them to the opposite places. This time, they are going find out how many they need to subtract to reach the given solution. Work through an example together. For example 8 – ☐ = 2. Find 8 on

Chapter 2 Addition and subtraction within 10

Unit 2.11 Practice Book 1A, pages 64–65

the number line and draw an arched arrow to 2. Count how many to get from 8 to 2 and complete the number sentence 8 – 6 = 2.

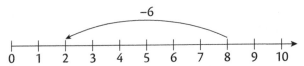

- Give pupils number lines to help them solve all the addition and subtraction number sentences in Question 3.

Same-day intervention

- Revisit the floor number-line activity in Question 2, but this time explore subtracting from 10.
- Pupils start at 10 and aim to reach 0. If they would travel beyond 0, they must miss that turn.
- Ask pupils to help you record their jumps along the number line and the matching number sentences. Encourage some pupils to take over the recording.
- Discuss how the two games are the same and how they are different. In the first game, pupils were adding. In the second they were taking away.

Same-day enrichment

- Provide pupils with some number lines with curved lines from one number to another already drawn on, for example:

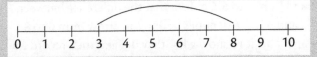

Ensure that there is no arrow head on either end of the line and no label.

- Ask pupils to think about what the calculation might have been. Ask them to record at least two number sentences for each number line. If necessary, point out that the curved linking line does not have any direction indicated on it.

Question 4

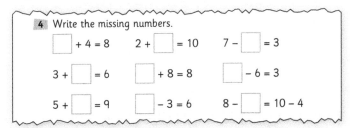

What learning will pupils have achieved at the conclusion of Question 4?

- Pupils will have explored finding a missing number in an addition or subtraction sentence.
- Pupils will have been introduced to using a number line to find the difference between two numbers.

Activities for whole-class instruction

- Show pupils an addition number sentence with a missing first addend, for example ☐ + 3 = 8.
- Read this as *Something add 3 equals 8*. Pupils should draw a jump of 3 that ends on 8. Can pupils see why subtracting 3 from 8 is the correct calculation – because it clearly shows where the jump begins and ends. If the number sentence tells us that the size of the jump was 3 and it ended at 8, the number line shows that the start number must have been 5. Subtracting the size of the jump from the end number gives us the start number. Show pupils how to draw a jump of 3 back from 8.

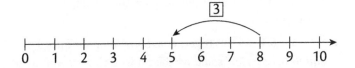

- Rewrite the number sentence using 2 as the second addend: ☐ + 2 = 8. Work through this together.
- Give pupils three more addition number sentences with missing first addends and ask them to solve these using a number line.
- Move on to an addition with the second addend missing, for example 4 + ☐ = 9.
- Read this as *4 add something equals 9*. Remind pupils that the number sentence tells them where the calculation begins, at 5. What they don't know is how many more are needed to equal 9. Ask pupils to look at the number line,

Chapter 2 Addition and subtraction within 10

Unit 2.11 Practice Book 1A, pages 64–65

find 4 and draw a jump to 9. Ask pupils to count along the jump to find out how many more must be added to 4 to reach 9. Complete the number sentence, 4 + 5 = 9.

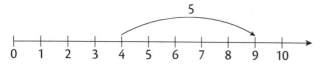

- Give pupils three addition number sentences with missing second addends and ask them to solve these using a number line.
- By counting along the number line, pupils can see that 9 was the whole, 9 − 4 = 5.
- Give pupils three subtraction number sentences with missing wholes and ask them to solve these using a number line.
- Alternatively, one of the parts could be missing, for example 8 − ☐ = 2. Read this together as *8 subtract something equals* 2.
- Remind pupils that they know that the whole is 8 and a part is 2, so to find the other part, they need to find the difference between 8 and 2. By drawing an arrow from 2 to 8, or 8 to 2, they can count along the number line to find out the difference. Draw a double-ended arrow and show pupils that they can count forward from 2 to 8 or back from 8 to 2, the difference is the same.

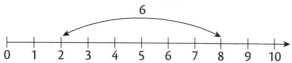

- Give pupils three subtraction number sentences with missing parts and ask them to solve these using a number line.
- ☐ − 3 = 6. Read this together as *Something subtract 3 equals* 6. Ask: *How could we solve this? Let's start with what we know. Do we know what the whole and parts are? What do we know? What don't we know? Can you tell me the two parts but not the whole?*
- One way to find the unknown whole is to add the two parts together. This would be one way to solve the problem – to think about the parts and whole.
- Returning to the number line, ask pupils to draw the jump of 3 back to 6. Where did it start? Show pupils the two parts (3 and 6) on the number line. Point out the similarity with a bar model.

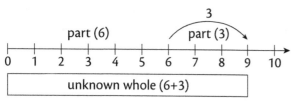

- Pupils have now practised solving addition and subtraction number sentences with a missing number in any position, using a number line for support. Question 4 will give them further practice in identifying and applying the most appropriate strategy.

Same-day intervention

- Show pupils the number sentence 3 + ☐ = 7. Check that pupils recognise that they know one part (3) and the whole (7), so they need to find out how many more to add to 3 to make 7.
- Ask them to link three cubes together and align them with their number line. Ask them to now put a finger on 7 and count from 3 to 7, to find out how many more cubes are needed to make 7.
- Ask pupils to draw an arrow from 3 to 7 to show what they did, label it and record the number sentence 3 + 4 = 7.
- Give pupils three more number sentences to practise finding the missing addend, using cubes and a number line.
- Show pupils the number sentence 8 − ☐ = 6. Check that pupils recognise they know the whole (8) and one of the parts (6), so they need to find out how many to subtract from 8 to leave 6.
- Ask pupils to get cubes, join them together and align them with their number line. Ask them to now put a finger on 8, the whole, to see how many cubes were subtracted to leave 8. Ask pupils to count from 6 to 8 and from 8 to 6. Confirm that the difference between 8 and 6 is 2. Ask pupils to draw an arrow from 6 to 8 and label it to show what they did and then record the number sentence 8 − 2 = 6.
- Give pupils three more number sentences to practise finding the missing subtrahend using cubes and a number line.

Chapter 2 Addition and subtraction within 10

Unit 2.11 Practice Book 1A, pages 64–65

Same-day enrichment

- Show pupils the last part of Question 4: 8 − ☐ = 10 − 4.
- Using a number line, draw a jump to represent 10 − 4 (from 10 to 6) to show that 10 − 4 = 6. Remind pupils that the equals sign means that both sides of the number sentence must have the same value, 6. So 8 − ☐ must equal 6. The jump from 8 must also end at 6 or the calculation would not balance. Draw in a jump from 8 to 6 and write the completed number sentence 8 − 2 = 10 − 4.
- Challenge pupils to make up three similar statements. They should first record an addition or subtraction number sentence and solve it and then create a second addition or subtraction number sentence with the same value (sum or difference). On the number line, this means that the solution to both will be the same number, so both jumps will meet on the same number.

Challenge and extension question

Question 5

5 Look at each number sentence. Draw it on the number line and find the answer.

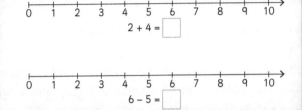

2 + 4 =

6 − 5 =

This question focuses on pupils using a number line to solve an addition and a subtraction number sentence. Encourage pupils to work from left to right along the number sentence. For 2 + 4, they highlight 2, count 4 numbers along the number line drawing in the jump of 4, landing on 6. So 2 + 4 = 6. Pupils may be able to solve each number sentence mentally. Can they visualise a number line to 10? A mental number line will be a vital tool for calculation.

Chapter 2 Addition and subtraction within 10

Unit 2.12
Games of number 10

Conceptual context

This unit focuses initially on pairs of numbers to make 10. After linking the numbers in pairs, they are also expressed as number sentences.

Pupils have seen and created many of these number sentences before. Removing the pairs of numbers for 10 from a number sentence helps pupils to link the pairs and begin to recall them.

This unit builds on previous units by exploring missing numbers in a range of positions in both addition and subtraction number sentences. This gives pupils the opportunity to revisit known strategies and explore how to use these for a particular purpose. It also revisits and extends equivalence, revisiting pairs of additions with equivalent value and extending to pairs of subtractions with equivalent value. This is then further extended to include inequalities between pairs of additions.

Learning pupils will have achieved at the end of the unit

- Pupils will have revisited working systematically to find all possibilities and be able to explain how they know that they have found all the addition bonds for 10 (Q1, Q2, Q3)
- Pupils will have found and begun to recall all the pairs of numbers that add to 10 (Q1, Q2, Q3)
- Pupils will have recognised that the equivalence of number bonds can be extended from addition to subtraction (Q4)
- Pupils will have explored recording equivalent relationships in a number sentence format (Q4)
- Understanding of the relationship between addition and subtraction will have deepened through continued use of bar models (Q5)
- The use of bar models to add and subtract will be becoming more fluent (Q5)
- Pupils will have revisited comparing quantities, recording the comparisons with <, > and = (Q6)
- Pupils will have applied learning from Chapter 1 about <, > and = to express more sophisticated numerical relationships (Q6)

Resources

ten-frames; counters; balance scales; cubes; 0–10 number lines; beans

Vocabulary

ten-frame, systematically, rotate, missing number, addition, addend, sum, subtraction, minuend, subtrahend, difference, equals, equivalent, inequality, is greater than (>), is less than (<)

Chapter 2 Addition and subtraction within 10

Unit 2.12 Practice Book 1A, pages 66–67

Questions 1–3

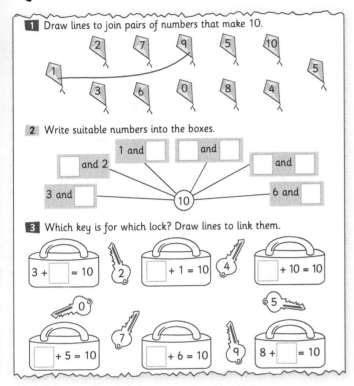

What learning will pupils have achieved at the conclusion of Questions 1–3?

- Pupils will have revisited working systematically to find all possibilities and be able to explain how they know that they have found all the addition bonds for 10.
- Pupils will have found and begun to recall all the pairs of numbers that add to 10.

Activities for whole-class instruction

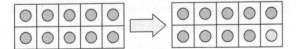

- Give pupils a ten-frame and ten double-sided counters, or 20 counters, ten each of two different colours. Ask pupils to place a counter on each space in the ten-frame, all the same colour. Count together to confirm that there are ten counters. Record the number sentence to match what the ten-frame shows: 10 + 0 = 10.
- Ask pupils to turn over (or swap for a different colour) one counter in the bottom right-hand corner. Ask pupils to say what they can see: nine counters in one colour and one in a different colour. Record the matching number sentence: 9 + 1 = 10.
- Ask pupils to turn over the next counter in the bottom row so that the two in the bottom right-hand corner are a different colour and again say what they can see: *Eight counters in one colour and two in another colour, 8 + 2 = 10.*
- Continue in the same way, changing the bottom row and then the top row from the right to left, until all ten counters have changed to the second colour and pupils have recorded all the number bonds for 10. Explain that by working systematically, they have found all the possible ways of adding two numbers to equal 10.
- Ask pupils to complete Questions 1–3 in the Practice Book.

Same-day intervention

- Carry out the same activity as above, but use a new ten-frame for each change in counter, building and displaying all 11 arrangements. Record the matching addition number sentence alongside each ten-frame as 10 ○ + 0 ○ = 10 counters, 10 + 0 = 10, 9 ○ + 1 ○ = 10 counters, 9 + 1 = 10 and so on.
- Verbally link the visual layout with the number sentence that includes counters. Then link that number sentence with the one without counters, saying that pupils could do exactly the same thing with something other than counters.
- Work steadily, taking time when a counter is changed to ensure that pupils can see the patterns involved. Continue until all the possible combinations are on display. Show pupils that they have used all the numbers from 0 to 10 as the first addend and as the second addend. There are no more numbers to use, so they have found all the possible ways of adding two numbers to equal 10.

Same-day enrichment

- Give pupils a ten-frame and ten double-sided counters, or 20 counters, ten each of two different colours. Ask pupils to place a counter on each space in the ten-frame, all the same colour. Count together to confirm that there are ten counters. Record the number sentence to match what the ten-frame shows, 10 + 0 = 10.
- Ask pupils to turn over (or swap for a different colour) one counter in the bottom right-hand corner. Ask pupils to say what they can see: nine counters in one colour and one in a different colour. Record the matching number sentence: 9 + 1 = 10. Rotate the ten-frame through 180 degrees and ask pupils if they can also see 1 + 9 = 10. Rotate the ten-frame back to the original orientation.

Chapter 2 Addition and subtraction within 10

Unit 2.12   Practice Book 1A, pages 66–67

- Ask pupils to turn over the next counter in the bottom row so that the two in the bottom right-hand corner are a different colour and again say what they can see: *Eight counters in one colour and two in another colour,* 8 + 2 = 10. Again, rotate the ten-frame through 180 degrees and ask pupils if they can also see 2 + 8 = 10.
- Continue in the same way until there are five counters, rotating each arrangement by 180 degrees and back again to see all the number pairs.
- Ask pupils to look at their pairs of additions and list all the pairs of numbers that make 10: 0, 10; 10, 0; 1, 9; 9, 1; 2, 8; 8, 2; 3, 7; 7, 3; 4, 6; 6, 4; 5, 5.

- Ask pupils to stand both sticks upright and compare them. They are the same height. Ask pupils to lay the sticks down side by side and compare them; they are the same length. Remind pupils that they made the sticks to represent the number sentences 4 + 2 = 6 and 5 + 1 = 6. Since both number sentences have a sum of 6, we can write them as 4 + 2 = 5 + 1.
- Place four cubes and two cubes in one pan with five cubes and one cube in the other pan. Check that pupils recognise that the scales balance and the matching number sentence is 4 + 2 = 5 + 1.
- Give small groups a balance scales and some cubes and ask them to create and record some balanced addition number sentences of their own.
- Show pupils two subtraction calculations, for example 8 − 2 = ☐ and 10 − 4 = ☐. Ask pupils to model each calculation with cubes and compare the differences. Both differences are 6. Ask pupils to check this by placing the differences in cubes in the two pans. The pans balance so 8 − 2 = 6 and 10 − 4 = 6. The two subtractions are equivalent because they have the same difference, so 8 − 2 = 10 − 4.
- Ask pupils to complete Question 4 in the Practice Book.

Question 4

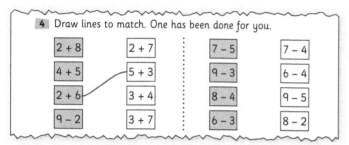

What learning will pupils have achieved at the conclusion of Question 4?

- Pupils will have recognised that the equivalence of number bonds can be extended from addition to subtraction.
- Pupils will have explored recording equivalent relationships in a number sentence format.

Activities for whole-class instruction

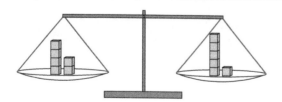

- Ask pupils to remind you what the = sign means. Confirm that both sides of the equals sign have the same value, though they may look different. Remind them that this is best illustrated by a balance scales. With the same amount in both pans, the pans are both at the same level. It is balanced.
- Show pupils the number sentences 4 + 2 = 6 and 5 + 1 = 6. Ask them to make a stick of six cubes, with four cubes in one colour and two cubes in a different colour. Now ask pupils to make another stick of six cubes using different colours: five cubes in one colour and one in another.

Same-day intervention

- Show pupils the number sentences 4 + 2 = 6 and 5 + 1 = 6 again. Ask them to make a stick of six cubes, with four cubes in one colour and two cubes in a different colour. Now ask pupils to make another stick of six cubes using different colours, five cubes in one colour and one in another.
- Ask pupils to stand both sticks upright and compare them. They are the same height. Ask pupils to lay the sticks down side by side and compare them; they are the same length. Remind pupils that they made the sticks to represent the number sentences 4 + 2 = 6 and 5 + 1 = 6. Since both number sentences have a sum of 6, we can write them as 4 + 2 = 5 + 1.
- Ask pupils to find a different way to make a stick of 6 using cubes in two different colours, so that both parts have the same value.
- Ask pupils to lay all the sticks side by side and stand them all up as towers to check that all three are the same length and height.
- Now ask pupils to make a stick of eight cubes all the same colour. Ask them to break off two cubes and record the matching number statement: 8 − 2 = 6. Their fourth stick is now also the same length and height as the other three sticks. Record 8 − 2 = ☐ + ☐. Ask pupils to choose one of the three addition sticks and complete the number statement accordingly.

Chapter 2 Addition and subtraction within 10

Unit 2.12   Practice Book 1A, pages 66–67

- Check that pupils recognise it does not matter if there is an addition on one side of the equals sign, additions both sides or subtractions both sides. As long as the calculations have the same value, they can be joined with the equals sign, because they are equivalent in value even though they do not look the same.

- Challenge pupils to find at least two more solutions. For each solution, they should also record the calculations used in separate number sentences.

Question 5

5 Fill in the boxes with suitable numbers.

5 + ☐ = 10   10 − ☐ = 0   9 + ☐ = 10   10 − ☐ = 1

3 + ☐ = 10   ☐ + 2 = 10   10 = 7 + ☐   ☐ − 5 = 5

☐ + 8 = 10   3 + ☐ = 7   4 + ☐ = 8   ☐ − ☐ = 3

What learning will pupils have achieved at the conclusion of Question 5?

- Understanding of the relationship between addition and subtraction will have deepened through continued use of bar models.
- The use of bar models to add and subtract will be becoming more fluent.

Same-day enrichment

- Show pupils the following addition puzzle.

	+		→	
+				+
↓				↓
	+		→	10

- Ask pupils what they notice. Show pupils that the grid has two routes to reach 10 in the bottom right-hand corner. Along the top and down the right-hand side is ☐ + ☐ = ☐ then ☐ + ☐ = 10. The two number sentences are linked with the sum of the first number sentence becoming the first addend of the second number sentence. For example: 2 + 4 = 6, 6 + 4 = 10. The second route is down the left-hand side and along the bottom. It also has ☐ + ☐ = ☐ then ☐ + ☐ = 10.

- If necessary, draw attention to the fact that the number in the bottom right-hand corner is 10, so the two calculations leading to 10 must be number bonds for 10. If two of those are entered into the grid, pupils can work backwards to find a starting number.

- Explain that an arrow is used to show the pathway to the next calculation. The equals sign cannot be used because both calculations will not have the same value.

- Work together to find a solution, for example:

3	+	5	→	8
+				+
6				2
↓				↓
9	+	1	→	10

Each calculation could be written separately: 3 + 5 = 8, 8 + 2 = 10 and 3 + 6 = 9 and 9 + 1 = 10. Show that the calculations are linked but not equal, so they cannot be connected using the equals sign. 3 + 5 = 8 but is not equivalent to 8 + 2.

Activities for whole-class instruction

- Remind pupils of the usual structure of a number sentence:

  ☐ ◯ ☐ = ☐

- To complete a number sentence, each of the square boxes needs a number and the circular box needs either the addition + or the subtraction − sign. A number sentence is a representation of a part–whole relationship and so is the bar model. So a missing number in a number sentence is an unknown number in a bar model.

whole	
part	part

- Show pupils the number sentence ☐ + 3 = 9. Ask them to label the bar model with the known quantities. It does not matter which part is labelled with 3. If the number sentence had been 3 + ☐ = 9, they could have labelled the bar model in the same way.

9	
part	3

- Ask pupils to either make the matching bar model using cubes or to use a number line to find the missing part.

Chapter 2 Addition and subtraction within 10

Unit 2.12   Practice Book 1A, pages 66–67

- Remind pupils that a part could also be missing from a subtraction number sentence. Show pupils the number sentence 9 – ☐ = 3 and label the bar model.

9	
part	3

- Check that pupils recognise they have drawn the same bar model. Just as with addition, they know the whole and one part and can use cubes in a bar model or a number line to help find the missing part.
- Finally, show pupils the number sentence ☐ – 6 = 3 and ask them to label a bar model with the known quantities. This makes it clear that the whole is unknown.

whole	
6	3

- Ask pupils to either make the matching bar model using cubes or to use a number line to find the unknown whole.
- Explain that drawing or visualising a bar model for the number sentence will help pupils identify whether it is a part or the whole that is missing.
- Ask pupils to complete Question 5 in the Practice Book.

Same-day intervention

- Give pupils some beans and a number line. Explain that there is often a missing number in an addition number sentence and they can use a number line to help them to find it. Show pupils the number sentence 9 + ☐ = 10. Read this together as 9 add something equals 10. Ask them to count out nine beans and match a bean to each number on the number line. Count along the line as they match. Check that pupils have matched to 9. Remind pupils that they took nine beans and they have counted and matched along the number line to 9.
- Ask pupils to look at their number line and count how many more beans they need to reach 10. Agree that they need 1 more. Ask pupils to add another bean to their number line and count along the number line to confirm that they have 10. Complete the number sentence 9 + 1 = 10.
- Remind pupils that addends can be in any order. Show pupils the number sentence ☐ + 3 = 8. Agree that one part is 3, the other part is unknown but they know the whole is 8. Ask pupils to take three beans and match them to the number line as they did before. Count along the line together to confirm that they have matched to 3.

Ask pupils to place a finger on 8, reminding them that this is the whole. Ask pupils to look at their number line and count from 3 to 8 to find out how many more beans they need to reach 8. Agree that 5 more beans would take them to 8. Ask pupils to count out five beans and add them to their number line to confirm that they have reached the whole, 8.

- Give pupils two more number sentences with missing addends to complete.
- Explain that there is often a missing number in a subtraction number sentence too. Show them the number sentence 6 – ☐ = 2. Read this together as 6 subtract something equals 2. Ask them to count out six beans and match them to their number line. Count along the number line together to check that everyone has 6 and the last number they matched was 6.
- Ask pupils to point to 2 on the number line. Remind them that the number sentence was 6 subtract something equals 2, so what they need to find out is how many beans they need to remove to have 2 left. Ask pupils to remove the beans between 2 and 6. Count them together and confirm that there are 4, so 6 – 4 = 2.
- Show pupils the number sentence ☐ – 4 = 3. Read this together as something subtract 4 equals 3. Ask pupils to count out three beans and place them on their number line. Remind them that 4 was subtracted from something to leave 3, so if they put the 4 back, they can see what the something was. Ask pupils to take four beans and match them to their number line. Confirm that they have reached 7, so 7 – 4 = 3. Ask pupils to remove four beans to check.

Same-day enrichment

- Show pupils a partly completed bar model with the value of a part unknown. Ask them what the missing number sentence could have been.

8	
part	2

- Check that pupils identify all four possible number sentences: 2 + ☐ = 8, ☐ + 2 = 8, 8 – ☐ = 2 and 8 – 2 = ☐. Ask pupils to solve one of the number sentences in any way they choose and use their solution to complete the other three.

Chapter 2 Addition and subtraction within 10

Unit 2.12 Practice Book 1A, pages 66–67

- Show pupils another partly completed bar model with the value of the whole unknown. Ask them to record the possible missing number sentences and select one to solve in any way they choose. They should then use their solution to complete the other three number sentences.

whole	
7	2

Question 6

6 Write >, < or = in each ◯.

7 + 3 ◯ 8   5 + 4 ◯ 7   2 + 7 ◯ 5 + 3
4 – 2 ◯ 6   9 – 3 ◯ 6   5 + 3 ◯ 9 + 0
2 + 8 ◯ 10   8 – 5 ◯ 10   6 + 4 ◯ 5 + 5

What learning will pupils have achieved at the conclusion of Question 6?

- Pupils will have revisited comparing quantities, recording the comparisons with <, > and =.
- Pupils will have applied learning from Chapter 1 about <, > and = to express more sophisticated numerical relationships.

Activities for whole-class instruction

Inequalities

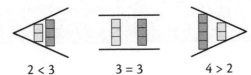

2 < 3   3 = 3   4 > 2

- Remind pupils that they told you that the equals sign means that both sides of the sign have the same value, so for 4 + 3 = 6 + 1, both sides have a sum of 7.
- Explain that if the two sides of the equals sign are not equal, we need a way to show it. Using a balance scales, put sticks of three and four cubes in one side of the balance. Ask pupils what should go in the other side pan to make it balance. Agree it should be 7 but place a stick of 6 in the pan. The pan with 7 in is clearly heavier than the pan with 6 in. Show pupils that this can be recorded as 4 + 3 is greater than 6, 4 + 3 > 6, and as 6 is less than 4 + 3, 6 < 4 + 3.
- Draw examples of the signs as in the illustration above. Explain that the pointed end of the sign is clearly the smaller end and the open end is clearly the larger end, so < is read as 'is smaller than' and > is read as 'is greater than'.
- Write up some paired statements, for example 3 < 4, 4 > 3 to show pupils that when comparing two quantities, either quantity could come first because the sign can point in either direction. Return to the balance scales and swap the contents of the two pans over. The recording will not change, 4 + 3 is greater than 6, 4 + 3 > 6, and as 6 is less than 4 + 3, 6 < 4 + 3.
- Ask pupils to work in groups. They should first explore and record some simple inequalities such as 4 < 6 and 6 > 4 and then some with an addition in one pan, for example 3 + 2 < 7, 7 > 3 + 2, and finally an addition in both pans: 6 + 2 < 4 + 5, 4 + 5 > 6 + 2.
- Give pupils a copy of the 'is less than' and 'is greater than' illustration to refer to when working through Question 6.
- Ask pupils to complete Question 6 in the Practice Book.

Same-day intervention

- Remind pupils what the equals sign means: that both sides of the equals sign have the same value.
- Ask pupils to put a tower of two cubes and a single cube on one side of the balance and a tower of three cubes on the other side. Once pupils have seen that they balance, take the cubes off the pans, laying them out to show the number sentence:

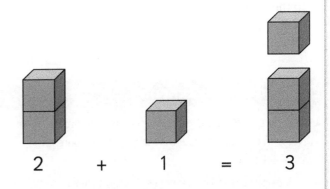

2 + 1 = 3

- Now ask pupils to make a tower of four cubes and a tower of five cubes. Ask them to compare the towers. Say: *4 is less than 5. It is shorter than 4.* Then ask pupils to repeat this with you. Then say: *5 is greater than 4. It is taller than 4.* Again, ask pupils to repeat this with you.
- Put one tower in each pan of the scales. Check that pupils recognise that the scales do not balance. Remind pupils that they just noticed that 5 is greater than 4, so the pan with 5 in is lower than that with 4 in, because

Chapter 2 Addition and subtraction within 10

Unit 2.12   Practice Book 1A, pages 66–67

five cubes are heavier than four cubes. Show pupils how to record 5 is greater than 4 with the matching inequality sign: 5 > 4.

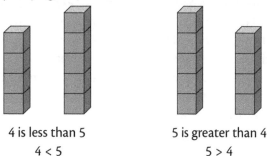

4 is less than 5   5 is greater than 4
4 < 5   5 > 4

- Then remind pupils that when they compared the two towers, they also noticed that 4 is less than 5, so the pan with 4 in it is higher than that with 5 in, because four cubes are lighter than five cubes. Show pupils how to record '4 is less than 5' with the matching inequality sign: 4 < 5.
- Show pupils the two inequalities number statements. Read them together and then look at the signs. Confirm that the pointed end always points to the smaller number. Show pupils a copy of the illustration above to confirm that the smaller quantity fits near the point and the greater quantity pushes the end of the inequality sign open.
- Repeat with towers of 6 and 4. After comparing the two quantities, split the tower of 6 into 4 and 2. Check that pupils can see that it makes no difference to the inequalities, 4 is less than 6, 4 < 6 and 6 is greater than 4, 6 > 4. Even if the 6 is split into 4 and 2, 4 < 4 + 2 and 4 + 2 > 4.

Same-day enrichment

- Remind pupils that they have used the inequalities signs 'is less than' < and 'is greater than' > to compare two numbers before.
- Give pupils a 0–10 number line. Remind them that < means 'is less than', for example 3 is less than 5, 3 < 5. Ask pupils to look at 3 and 5 on the number line. 3 is to the left of 5 on the number line, because the number line shows the numbers in order.

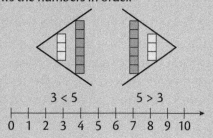

- Remind pupils that > means 'is greater than', so 5 > 3. Ask pupils to look at 5 and 3 on the number line. The 5 is further along the line to the right because it is greater in value than 3.
- Remind pupils that both sides of the equals sign have the same value. If they don't, then the equals sign cannot be used. Instead we can use an inequality sign, to show which side has the greater or lesser value.
- Challenge pupils to record pairs of unequal number sentences using both the 'is greater than' and 'is less than' inequalities signs.

Challenge and extension question

Question 7

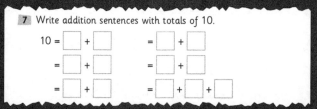

7 Write addition sentences with totals of 10.

This question focuses on addition bonds for 10. Pupils have rarely been presented with the sum at the beginning of the number sentence. They have, however, seen two addends at the beginning of a number sentence with further groups of two addends following, so they should quickly realise what is needed. Pupils may use a pattern of number pairs to complete the extended sentence. However, the final part of the number sentence has three empty number boxes rather than two.

Chapter 2 Addition and subtraction within 10

Unit 2.13
Adding three numbers

Conceptual context

This unit focuses on extending addition from adding two quantities to adding three. The process is the same as it was when there were two addends so previous approaches are revisited and extended.

Pupils are initially shown a bar model with three parts and a whole so that they have a clear visual image of what the extension means. They then explore solving three-part additions using the familiar approaches of ten-frames, sticks of cubes and the number line.

Pupils consolidate their understanding through recording a story in an addition number sentence, adding three numbers.

Learning pupils will have achieved at the end of the unit

- Pupils will have used the bar model to represent wholes with three parts (Q1)
- Pupils will have explored adding three numbers together using a variety of representations (Q1, Q2, Q3, Q4)
- Pupils will have recognised that the three addends can be recorded in any order in an addition number sentence (Q1, Q2, Q3)
- Starting an addition with the largest addend will have been explored and demonstrated to be a more efficient approach (Q3)
- Pupils will have generated and solved an addition number sentence from a story (Q4)

Resources

mini whiteboards; interlocking cubes; beans; counters; ten-frames; 0–10 number line; sets of digit cards (two sets of 0–3 and a 4 per pair)

Vocabulary

bar model, ten-frame, addend, sum, number line

Chapter 2 Addition and subtraction within 10

Unit 2.13 Practice Book 1A, pages 68–69

Question 1

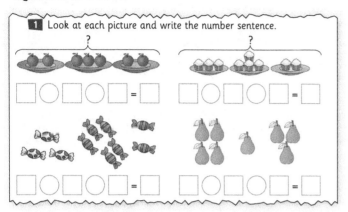

What learning will pupils have achieved at the conclusion of Question 1?

- Pupils will have used the bar model to represent wholes with three parts.
- Pupils will have explored adding three numbers together using a variety or representations.
- Pupils will have recognised that the three addends can be recorded in any order in an addition number sentence.

Activities for whole-class instruction

- Draw a bar model with three parts.

whole		
part	part	part

- Ask pupils to explain how this is the same and how it is different from the bar models they have seen before. Check that pupils recognise that it has three parts instead of two, so the whole has the same value as all three parts.
- Ask pupils to use different coloured cubes to make sticks of 2, 3 and 4 cubes. Show pupils how to lay the three sticks in a row, just like the three parts in the bar model diagram.
- Now ask them to make another stick in a different colour, the same size as the whole stick. Show pupils this bar model and ask pupils if theirs is the same, though their colours may be different. Confirm that provided they have a whole of 9 and three parts of 2, 3 and 4, their bar model is the same.

			9			
	2	3			4	

- Ask pupils to tell you a matching number sentence for your bar model. Agree that we normally read from left to right, so 2 + 3 + 4 = 9 would be correct. Ask pupils if they have a different number sentence. Record these alongside your number sentence and confirm that they simply have the addends in a different order; their sentences are still correct.
- Record the number sentence addend + addend + addend = sum and confirm that the bar model shows that the addends can be in any order, the sum will be the same. Ask pupils to reorder their part sticks to check.
- Ask pupils to complete Question 1 in the Practice Book.

Same-day intervention

- Tell pupils that they have had lots of practice at adding two numbers together, so now they are going to look at adding three numbers together.
- Give pupils a ten-frame and beans, counters or cubes in three different colours.
- Ask pupils to take two beans and place them on their ten-frame, working along the top row. Ask pupils how many beans they have and record this as the beginning of a number sentence, 2.
- Ask pupils to take three different coloured beans and add them to their ten-frame, continuing along the top row. Ask pupils how many beans they have now. Some pupils may recognise that one complete row of the ten-frame is 5 and be able to answer very quickly. Count the beans together and show that the number sentence to represent what they done so far is 2 + 3. Confirm that you could write 2 + 3 = 5, but you haven't finished yet, there is more to add.
- Ask pupils to take three beans in the final colour and place them on the ten-frame, in the second row. Ask pupils how many beans they have now. Some pupils may recognise that the two empty spaces mean that there must be eight beans on the ten-frame and be able to answer very quickly.
- Ask pupils how you could complete the number sentence. Talk through, recording 2 + 3 + 3 = 8. Read the number sentence together, pointing to the relevant line of beans.
- Ask pupils to return the beans to their piles and work through another calculation adding three numbers.

Chapter 2 Addition and subtraction within 10

Unit 2.13 Practice Book 1A, pages 68–69

Same-day enrichment

- Draw a bar model with three parts.

whole		
part	part	part

- Ask pupils to explain how this is the same and how it is different from the bar models they have seen before. Check that pupils recognise that it has three parts instead of two, so the whole has the same value as all three parts.
- Show pupils three small piles of beans, with two, three and four beans. Explain that the piles of beans are parts, the whole is all the beans. Remind pupils that they know how to record a number sentence and ask them to write a number sentence for the beans.
- Pupils may record something like 2 + 3 = 5, 5 + 4 = 9. Agree that this shows their thinking, that they added two of the quantities together first, then added the third quantity. Show them that there is a shorter way: 2 + 3 + 4 = 9.
- Give pupils the layout ☐ + ☐ + ☐ = △, two sets of digit cards 0–3 and a single 4 card. Ask pupils to shuffle the cards and turn over the first three cards to generate the numbers to go in the square boxes. The triangular box is for the sum and they can write in there. Challenge pupils to record three three-part additions and solve them in any way they choose.

Question 2

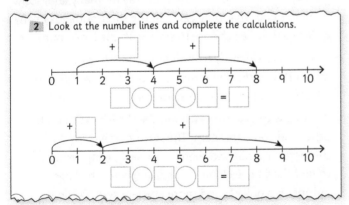

What learning will pupils have achieved at the conclusion of Question 2?

- Pupils will have explored adding three numbers together using a number line.
- The idea that the three addends can be recorded in any order in an addition number sentence will have been revisited.

Activities for whole-class instruction

- Remind pupils that when they added two numbers together on a number line, they first looked along the number line to find the value of the first addend and then drew a jump the size of the second addend to find the sum.
- Ask pupils how many jumps they will need to draw on a number line to add three quantities. Agree that they will need to draw two jumps.
- Ask pupils to use different coloured cubes to make sticks of two, two and five cubes. Show pupils the number sentence 2 + 2 + 5 = ☐. Ask pupils how they could show the first stick of two on the number line. Confirm that they can identify where 2 is on the number line, the matching number. Ask pupils to put the stick of two cubes to one side and pick up the second stick. Confirm that they need to draw in a jump of 2 along the number line, starting with 2. Ask pupils where their jump ends. Confirm it ends at 4. Pupils should label their jump and join their first two sticks together. Ask pupils to lay them near the number line and count along the stick, then along the number line, to confirm they have the same sum. Explain that they are just using two different representations together to help them understand what is happening.
- Ask pupils to pick up their third stick. Confirm that they need to draw a jump of 5 on the number line because they are adding 5 more. Ask pupils where the jump should start, at 4 because they have 4 already. After pupils have drawn and labelled their jump, ask them where this jump finished. Confirm that it finished at 9, because 2 + 2 + 5 = 9. Pupils should record the number sentence next to the number line.
- Ask pupils to join their sticks together and count the cubes to confirm that there are 9 altogether. Ask pupils to place their joined sticks close to their number line and point to the relevant parts of the stick as you read the number sentence together. Read the number sentence together again, this time pointing to the jumps on the number line.
- Ask pupils to complete Question 2 in the Practice Book.

Chapter 2 Addition and subtraction within 10

Unit 2.13 Practice Book 1A, pages 68–69

Same-day intervention

- Give pupils counters in three different colours and a number line.
- Ask pupils to take two counters in one colour and place them next to the numbers 1 and 2. Explain that they have reached 2 on the number line because they have two counters.
- Now ask pupils to take three counters in a different colour and place them along the number line, after 2 and with one counter for each number. Ask pupils to tell you how many counters they have altogether: five because the last counter they placed on the number line was at 5. Encourage pupils to count to check.
- Now ask pupils to take four counters in the final colour and place them along the number line, after 5 and with one counter for each number. Ask pupils to tell you how many counters they have altogether: 9 because the last counter they placed on the number line was at 9. Encourage pupils to count to check.
- Show pupils how they can record the calculation with a jump of 3 from 2 to 5 to represent the three counters added to five and a further jump of 4 from 5 to 9 to represent the four counters added to five. Write the matching number sentence for the calculation: 2 + 3 + 4 = 9.
- Check that pupils can see what each number in the calculation represents by comparing it with the counters and the number line.
- Ask pupils to work through Question 2.

Same-day enrichment

- Remind pupils that the calculation for the second number line in Question 2 was 0 + 2 + 7 = 9. The first jump started from 0 because the first addend was 0.
- Rearrange the number sentence to 2 + 0 + 7 = 9 and ask pupils to represent this on a number line.
- Ask pupils to explain how their number line represents 2 + 0 + 7 = 9 rather than 2 + 7 = 9. Share ideas for representing a jump of zero.
- Ask pupils to rearrange the number sentence again to 2 + 7 + 0 = 9 and show this on the number line to check that the representation of 0 works regardless of the position of the addend of 0.

Question 3

3 Complete the calculations.

1 + 3 + 3 = ☐   0 + 3 + 4 = ☐   8 + 0 + 2 = ☐
3 + 4 + 0 = ☐   2 + 2 + 1 = ☐   1 + 6 + 2 = ☐
2 + 1 + 6 = ☐   5 + 3 + 2 = ☐   2 + 7 + 1 = ☐
2 + 4 + 1 = ☐   4 + 2 + 4 = ☐   1 + 4 + 5 = ☐

What learning will pupils have achieved at the conclusion of Question 3?

- Pupils will have consolidated adding three numbers together using a number line or sticks of cubes.
- Starting an addition with the largest addend will have been explored and demonstrated to be a more efficient approach.

Activities for whole-class instruction

- Give pupils two number lines and explain that you are going to work through the addition 2 + 1 + 6 = ☐ together.
- With pupils, identify 2 (the first addend) and draw a jump from 2 to 3, labelling it + 1 for the second addend. Then draw in a jump of 6, from 3 to 9, for the third addend. Label the jump + 6 and identify the sum. Complete the number sentence 2 + 1 + 6 = 9. Remind pupils that this is what they did in Question 2.
- Compliment pupils on working quickly and accurately, but explain that there is a quicker way for many additions. Remind pupils that they can add the three addends in any order, so they can choose to start with the larger number.
- Ask pupils to identify 6 on their number line (one addend) then draw in jumps of 2 and 1 for the other two addends, again reaching the sum of 9.
- Ask pupils to place the two number lines next to each other and compare them. The second one started with the largest number, so there was less drawing needed to find the sum. Explain that starting with the largest number is a quicker and more efficient way to add because it gives you less work to do.
- Ask pupils to look at Question 3. Ask them to draw a ring around the largest number in each calculation and to begin their addition on the number line with that number.

Chapter 2 Addition and subtraction within 10

Unit 2.13 Practice Book 1A, pages 68–69

Same-day intervention

- Ask pupils to make sticks of two, one and 6 cubes. Explain that you would like pupils to join all the sticks together in the order given, to show 2 + 1 + 6 = ☐. Count along the whole stick to find the total of 9, 2 + 1 + 6 = 9.

- Explain that, instead of counting all the cubes, it is quicker to count on from the first number. Ask pupils to place a finger on the second of the cubes on the 2 stick. Remind them that this stick is made of two cubes. They know that so they don't need to count from the beginning again. Count on from 2 together to 9.

- Tell pupils that although that involved less counting and they still reached the same sum because the addends can be added in any order, there is an even better way.

- Ask pupils to turn their stick around as if the first stick they made was the 6 stick. Remind pupils that this is the largest of the three sticks, so if they start counting on from this one, there will be less to count. Remind them that there are six cubes in this part of the stick and ask them to put a finger on the last cube of the 6. Count on together from 6 to 9 together. Check that pupils recognise that there was less counting involved so it was a quicker way to find the sum.

- Ask pupils to make sticks of two, three and five cubes. Ask them which is the largest stick, the one to start with. Confirm that this is 5. 2 and 3 are close together in value so it does not matter which one comes next, as long as 5 is first. Count on from 5 to find the sum, 10. Record the number sentence: 5 + 3 + 2 = 10.

- Choose some of the calculations in Question 3 for pupils to solve in the same way.

Same-day enrichment

- Show pupils the following puzzle. Explain that every row of three numbers, every column of three numbers and even the diagonal arrangement of numbers (all indicated by arrows) have a sum of 10. Challenge pupils to find a solution.

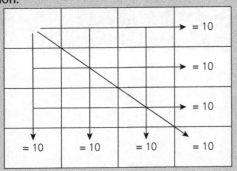

- Pupils may find it helpful to list some sets of three numbers that add to 10 first of all. They could use a ten-frame, cubes, a number line or something else to help them.

- Share some solutions and ask pupils to explain how they found their solution.

Question 4

4 Write the number sentences.

(a) 3 children were playing football in the playground. Another 2 children joined them. Then 5 more children joined the game. How many children played football?

Number sentence: _____

(b) Amina has 4 red pencils and 2 blue pencils. She has the same number of green pencils as blue ones. How many pencils does she have altogether?

Number sentence: _____

What learning will pupils have achieved at the conclusion of Question 4?

- Pupils will have generated and solved an addition number sentence from a story.

Activities for whole-class instruction

- Say to pupils that you are going to tell them some addition stories. Ask them to record a matching number sentence and then solve the number sentence in any way they choose. They could use cubes, a ten-frame, a number line or something else.

- Begin the story: *There were three children swimming in the swimming pool.* Pupils should record 3. *three more children arrived to join them.* Pupils should extend their recording to 3 + 3. Continue the story: *In a few minutes, another four children arrived.* Pupils should extend their recording to 3 + 3 + 4 = ☐. *How many children were swimming in the swimming pool?*

- Give pupils a few minutes to find the sum, then share some different ways of finding a solution.

- Tell a second story: *Shamal's dog had a litter of puppies. three of them were brown* (pupils should record 3) *and two of them were black* (pupils should extend their recording to 3 + 2). *There were also two white ones. How many puppies were in the litter?* Pupils should extend their recording to 3 + 2 + 2 = ☐.

- Challenge pupils to try a different method to find the sum.

137

Chapter 2 Addition and subtraction within 10

Unit 2.13 Practice Book 1A, pages 68–69

- Ask pupils to complete Question 4 in the Practice Book.

Same-day intervention

- Give pupils a ten-frame and some counters or small counting objects.
- Explain that you are going to tell them a story. You would like them to model as you tell it so that they can quickly find the sum.
- Begin with: *Sumi was blowing up balloons for a party. She managed to blow up three before she needed a rest.* Ask pupils to take three counters and represent the blown up balloons in the top row of the ten-frame.
- Continue the story: *After a rest, Sumi blew up three more balloons.* Ask pupils to take three more counters in a different colour and add them to their frame. Pupils may be able to tell you there are six balloons by recognising that there is one full row and 1 more. Alternatively, some pupils may be able to count on from the original three counters. Model both approaches to support pupils who are counting all.
- Finish the story: *Sumi blew up two more balloons.* Ask pupils how many balloons Sumi blew up altogether. Pupils should take two more counters in a different colour to add to their ten-frame. Some pupils may recognise the layout as 8 because there are two more spaces to reach 10, or they may count on from 6. Model both approaches.
- Ask pupils to look at their ten-frame and tell you what the number sentence was that they just solved. Record $3 + 3 + 2 = 8$.
- Ask pupils to clear their ten-frame and tell them another story for them to model and solve.

Same-day enrichment

- Give pupils two sets of 0–3 digit cards and a 4.
- Ask pupils to work in pairs. They shuffle the cards. The first pupil turns over one card at a time, telling a story as they do so without showing the cards to their partner. Their partner keeps track of the problem using cubes or a number line, telling their partner the sum. The storyteller then reveals the three cards and they check the calculation together before swapping roles.
- It may be useful to draw up a list of topics for the stories before pupils begin. This will help the stories to flow. Useful topics include children, birds, horses, balloons or other things that match pupils' interests.
- Make sure that pupils have the opportunity to tell at least two stories each. This will help pupils to understand what written problems are asking of them.

Challenge and extension question

Question 5

5 Think carefully and then write the answers in the boxes.

$2 + 3 + \square = 9$

$\square + 1 + 2 = 6$

$4 + \square + 4 = 10$

$\square + 3 + 5 = 9$

This question encourages pupils to explore finding a missing addend when they are adding three numbers together to find a given sum. In effect, pupils are having to add the two given numbers and then find out how many more are needed to reach the given sum. Pupils may recall and use some number bonds to find the missing number. For example, the first calculation indicates that pupils should add 2 and 3 (5). They may then recall that $5 + 4 = 9$ and be able to use this to complete the number sentence. The missing numbers are in different places and pupils will have to use inverse operations to find some missing addends.

Chapter 2 Addition and subtraction within 10

Unit 2.14
Subtracting three numbers

Conceptual context

This unit focuses on extending pupils' knowledge of subtraction. The process is a natural expansion of work with addition of three parts in the previous unit. In this unit, a whole of 10, made up of three parts, is the basis for subtracting two of those parts from the whole.

Previous approaches are revisited and extended. Pupils will work practically with counters and cubes, physically taking away each quantity to find the difference. They will also explore using pictorial approaches and the number line.

Pupils consolidate their understanding through recording a story in a subtraction number sentence, subtracting two smaller numbers from a number up to 10.

Learning pupils will have achieved at the end of the unit

- Pupils will have explored and consolidated subtracting two numbers from a number up to 10 using a variety of representations (Q1, Q2, Q3, Q4)
- Pupils will have recorded two subtractions in the same number sentence (Q1)
- Pupils will have explored how to record the subtraction of 0 on a number line (Q2)
- Pupils will have revisited the link between addition and subtraction through number pairs for 10 (Q3)
- Pupils will have generated and solved a subtraction number sentence from a story (Q4)

Resources

ten-frames; mini whiteboards; counters; interlocking cubes; counting objects; coloured pencils; paper copies of a bar model; 0–10 number lines; 0–10 number cards

Vocabulary

subtract, subtraction, minuend, subtrahend, difference, bar model, ten-frame, number line, number bond, number pair

Chapter 2 Addition and subtraction within 10

Unit 2.14 Practice Book 1A, pages 70–71

Question 1

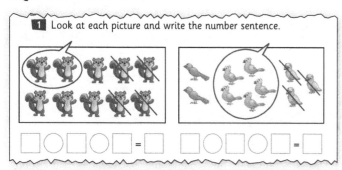

What learning will pupils have achieved at the conclusion of Question 1?

- Pupils will have explored subtracting two smaller numbers from 10 using a variety of representations.
- Pupils will have recorded two subtractions in the same number sentence.

Activities for whole-class instruction

- Give pupils a ten-frame and ten identical counters or objects. Ask pupils to put the counters in the frame, one in each space. Check that pupils are confident that they have ten, asking them to explain how they know.
- Explain that they have already explored adding three numbers together. Today, they will be exploring subtracting two numbers from 10.
- Remind pupils that they have already represented 10 on their ten-frame. Ask them to show you how to subtract 3 from their 10. Pupils may remove three counters in different ways. Suggest that if they remove counters beginning at the bottom right, they will find it easier to quickly 'read' how many counters are left without the need to count them. They now have 5 and 2 more left on the frame, 7 altogether.

- Ask pupils how they could record what they did in a number sentence. Confirm the subtraction number sentence as 10 − 3 = 7. Remind pupils that they are going to subtract a second amount, so they are not yet ready for the equals sign and the difference.
- Ask pupils to subtract four counters from what they have left. Check that everyone has three left and that they have removed them following the previous directions.

- Ask pupils how they could record this in a number sentence. Confirm this as 10 − 3 − 4 = 3. Confirm that subtracting two numbers is just the same as subtracting one number; the process is the same.
- If necessary, work through another example.
- Give pupils a copy of a ten-frame with ten circles or other shapes already drawn in. Ask pupils how they could use this to subtract 3 then 4 from 10, when they cannot remove the pictures. Discuss pupils' ideas and settle on using two different coloured pencils to cross out the pictures. Ask pupils to use one pencil to cross out three pictures. Check that pupils can see that 10 − 3 = 7 as before. Now they can cross out another four pictures, leaving three unmarked. Check that pupils recognise that they have carried out the same subtraction, using a slightly different method.
- Pupils are now ready to work through Question 1.

Same-day intervention

- Give pupils some cubes and ask them to make a stick of ten cubes, all the same colour. Count the cubes together to confirm that everyone has ten.
- Explain that you are going to subtract 3 from your stick of cubes and ask pupils how you could do it. Accept different explanations and say that you will break off a stick of three cubes so that you can quickly see that you have removed the correct number. Breaking off cubes one at a time would be slower and you may lose count.
- Ask pupils to do the same. Remind them that they started with ten cubes and subtracted 3, so they could record what they have done as 10 − 3 = 7. Ask pupils to check they have a stick of three and a stick of seven cubes, 10 altogether.
- Explain that you haven't finished yet. Ask pupils to put the stick of three cubes to one side and pick up the stick of seven cubes. Tell pupils that you would like to subtract another five cubes and ask them to show you how to do this. If necessary, model subtracting 5 by breaking off five cubes, leaving 2. Remind pupils that they recorded 10 − 3 = 7, but they have subtracted some more, so rub out the = 7 and complete the number sentence as 10 − 3 − 5 = 2.
- Model lining up all three sticks of cubes and removing the 3 stick then the 5 stick as you say: 10 − 3 − 5 = 2. Ask pupils to say the number sentence with you, carrying out the same actions.
- Repeat with another example.

Chapter 2 Addition and subtraction within 10

Unit 2.14 Practice Book 1A, pages 70–71

Same-day enrichment

- Draw a bar model with three parts. Remind pupils that they used this when they explored adding three numbers.

whole		
part	part	part

- Ask pupils to explain how this bar model also shows the part–whole relationship when subtracting two numbers from 10.
- Give pupils some photocopied outlines of this bar model and some scissors. Ask them to cut out the whole model, then cut off and discard the third part. Explain that the missing part is the difference between the whole and the known parts. It can also be thought of as how many more to make both bars the same.
- Ask pupils to make a stick of ten cubes – this is the whole. Ask pupils to make a stick of four cubes and a stick of two cubes and place them below the whole to represent the two parts. Ask pupils how they can find out how many more to make both bars equivalent in value.
- Try out the various suggestions. Some pupils may also recall and use the number bond 6 + 4 = 10.
- Show pupils how to count on from 6 to 10: 7, 8, 9, 10. Explain that they already have 6, so they need to say four more numbers to count to 10; four more are needed to reach 10. The difference between 6 and 10 is 4. Also show this difference on a number line, emphasising that the difference between 6 and 10 is 4.
- Record the number sentence 10 – 4 – 2 = 4.
- Give pupils some subtraction calculations to explore with cubes and using the bar model as a guide.

What learning will pupils have achieved at the conclusion of Question 2?

- Pupils will have revisited subtracting two smaller numbers from a number up to 10 using a variety of representations.
- Pupils will have explored how to record the subtraction of 0 on a number line.

Activities for whole-class instruction

- Remind pupils that when they first used a number line for subtraction, they started from the minuend and jumped back. If there are two subtrahends, there will be a second jump from where the first jump landed.
- Show pupils a 0–10 number line and ask them to talk you through how you would use it to subtract 5 from 9. Follow pupils' instructions, making mistakes where instructions are imprecise and asking pupils to check. Confirm that 9 – 5 = 4.

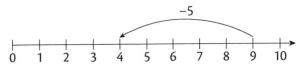

- Tell pupils that you need to subtract another 2 and ask pupils how you could do that. Confirm that you can draw a second arrow to show a jump of 2.

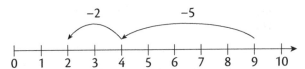

- Ask pupils to help you record the number sentence for what you did on the number line: 9 – 5 – 2 = 2.
- Give pupils two subtraction number sentences with two subtrahends to complete using a number line before asking them to complete Question 2.

Question 2

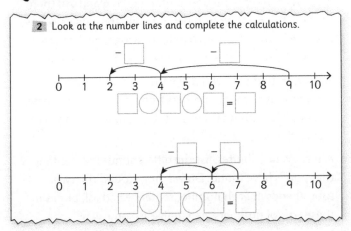

2 Look at the number lines and complete the calculations.

Same-day intervention

- Give pupils a number line with spacing the same size as the cubes they will be using and some cubes.
- Ask pupils to make a stick of eight cubes and to compare it with their number line. With one end of the cube stick at 0, the other end should be at 8, confirming that there are eight cubes in the stick. Pupils should adjust their stick until they have 8 if necessary.

Chapter 2 Addition and subtraction within 10 Unit 2.14   Practice Book 1A, pages 70–71

- Ask pupils to keep the stick where it is on the number line, but to break off, in other words subtract, three cubes. The stick of 8 has become a stick of 5 because three cubes been subtracted. Remind pupils how to record a jump of 3 to match the removed cubes, from 8 to 5 on the number line. Record the number sentence 8 − 3 = 5 together, checking that pupils recognise how each part is represented on their number line.

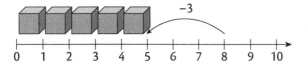

- Tell pupils that you would like them to subtract two more cubes from their stick. Again, pupils should keep their cube stick where it is and break off two more cubes from the 5 end.
- Ask pupils how they can record this subtraction of 2 on the number line. Confirm that they can draw another jump of 2, from 5 to 3, to show the subtraction and that they have three cubes left.

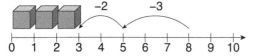

- Remind pupils that they recorded the number sentence 8 − 3 = 5 for the first jump. Show them how to use one number sentence for everything they did: 8 − 3 − 2 = 3. Read the subtraction sentence, pointing to the relevant parts of the number line. Ask pupils to repeat the number sentence with you, pointing to the relevant parts on their own number line.
- Work through another example together before asking pupils to complete Question 2.

Same-day enrichment

- Give each pair of pupils ten counters or 1 cm cubes and ask them to spread them out on the table, within easy reach but not too close together.
- Pupils take turns to place both hands over the counters, picking up some in each hand. Some pupils will find it easier to pick them up by sliding them to the edge of the table. Alternatively, ask one pupil to close their eyes while their partner picks some up, placing some in each hand.
- Their partner asks them to open one hand at a time and records those counters as a subtraction from 10 on the number line. Those in the second hand are a further subtraction, which can also be shown on the number line.

The two jumps show the number of counters removed from the set of ten with the second jump ending on the difference, the number of counters left on the table. For example, if a pupil picked up two counters in one hand and three in the other, there would be five left on the table and 10 − 2 − 3 = 5.

- Pupils swap roles until both have had at least three turns at both roles.

Question 3

> 3 Complete the subtraction calculations.
>
> 5 − 3 − 1 = ☐   6 − 3 − 3 = ☐   8 − 4 − 2 = ☐
> 6 − 4 − 0 = ☐   7 − 2 − 1 = ☐   9 − 3 − 5 = ☐
> 8 − 7 − 1 = ☐   10 − 3 − 4 = ☐   10 − 7 − 3 = ☐
> 8 − 2 − 2 = ☐   7 − 2 − 3 = ☐   6 − 3 − 1 = ☐
> 10 − 6 − 2 = ☐   6 − 5 − 1 = ☐   4 − 2 − 2 = ☐

What learning will pupils have achieved at the conclusion of Question 3?

- Pupils will have consolidated subtracting two numbers from a number up to 10 using a variety of representations.
- Pupils will have revisited the link between addition and subtraction through number pairs for 10.

Activities for whole-class instruction

- Give pupils ten cubes and a 0–10 number line.
- Ask them to make a stick of ten cubes and align it with their number line. Ask pupils to subtract 4 from their stick of 10 and record this on their number line with a jump of 4 (the subtrahend), from 10 (the minuend) to 6. Tell pupils that the second subtrahend is 6 and ask them to show this on their number line. Talk through drawing a further jump of 6 from 6 to 0, showing that there are 0 cubes left. Record the number sentence: 10 − 4 − 6 = 0.
- Tell pupils that you would like them to find different ways to subtract two numbers from 10 so that they have 0 cubes left. Remind them that the number sentence to record this is 10 − ☐ − ☐ = 0.
- After several minutes, list the different number sentences that pupils have found. Draw a ring around the two subtrahends in each number sentence and ask pupils if these pair of numbers are familiar in some way.

Chapter 2 Addition and subtraction within 10

Unit 2.14 Practice Book 1A, pages 70–71

- If pupils do not recognise the numbers, remind them that they are the number bonds for 10 that is pairs of two numbers that add together to make 10 (0, 10; 1, 9; 2, 8; 3, 7; 4, 6; 5, 5). Explain that what they have done is 'undo' those familiar additions by subtracting them from 10.
- Ask pupils to check that there are no pairs of numbers missing. Subtracting 10 and 0, or 0 and 10 is the most likely pair of numbers to be missed.
- Ask pupils to complete Question 3 in the Practice Book.

Same-day intervention

- Give pupils ten cubes to make into a stick. Ask them to count the cubes in their stick to confirm they have 10. Remove any spare cubes from the table.
- Explain that they are going to subtract two amounts from 10 to leave 0. Show them the number sentence 10 – ☐ – ☐ = 0 and explain that they can use this to record each solution that they find.
- Start pupils off by asking them to remove one cube from their stick of ten. Ask how many more they need to remove from their stick to have zero cubes left in the stick. If necessary, encourage pupils to count the cubes in the stick to confirm there are nine. Show pupils how to record this as 10 – 1 – 9 = 0. Read the number sentence together, modelling it with the whole stick of cubes on the table, then one cube removed, then the stick of nine cubes removed, leaving zero cubes on the table.
- Ask pupils to remake their stick of ten cubes. Work through removing two cubes then identifying that they need to remove eight to leave zero cubes. Record as before: 10 – 2 – 8 = 0.
- Support pupils to work systematically until they have found 10 – 5 – 5 = 0.
- If pupils want to continue, point out that they have already recorded subtracting 1, 2, 3, 4, 5, 6, 7, 8 and 9 from 10. If necessary, suggest they try subtracting 0 then 10 to complete their subtractions. List all the pairs of numbers that were subtracted and remind pupils that these are the number pairs that make 10 and there are only six different pairs.

Same-day enrichment

- Show pupils the following puzzle. Explain that every row of three numbers and every column of three numbers (all indicated by arrows) have a difference of 0. Challenge pupils to find a solution.

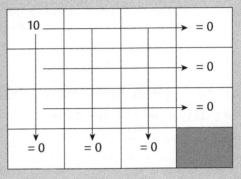

Example solution

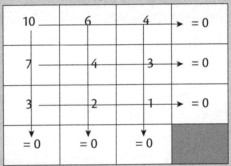

- Only the two subtractions in the top row and left-hand column are subtractions from 10. The other calculations use the subtrahends from those two calculations as minuends for further calculations.
- Share some solutions and ask pupils to explain how they found their solution.
- If a hint is needed, remind pupils to think about their number bonds for 10 as subtractions.
- Explain that there is one 'super solution' where even the diagonal has a difference of 0. Challenge pupils to find it.

Super solution

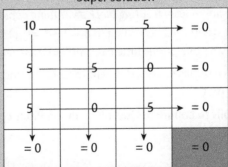

Chapter 2 Addition and subtraction within 10   Unit 2.14   Practice Book 1A, pages 70–71

Question 4

> 4 Read the problems and then write the number sentences.
>   (a) 9 birds were in the tree. 5 birds flew away. Then 2 more flew away. How many birds are still in the tree?
>
>   Number sentence: _____
>
>   (b) There are 9 balloons. 3 of them are blue. Another 4 are red. The rest are yellow. How many balloons are yellow?
>
>   Number sentence: _____

What learning will pupils have achieved at the conclusion of Question 4?

- Pupils will have generated and solved a subtraction number sentence from a story.

Activities for whole-class instruction

- Tell pupils that you are going to tell them some subtraction stories. Ask them to record a matching number sentence and then solve the number sentence in any way they choose. They could use cubes, a ten-frame, a number line or something else.
- Begin the story: *There were ten children swimming in the swimming pool*. Pupils should record 10. *Three children left to go home*. Pupils should extend their recording to 10 – 3. Continue the story: *A short while later, another four children went home*. Pupils should extend their recording to 10 – 3 – 4 = ☐. *How many children were left in the swimming pool?*
- Give pupils a few minutes to find the difference, then share some different ways of finding a solution.
- Tell a second story: *There were eight children swimming in the swimming pool* (pupils should record 8). Continue the story: *Two of them were wearing yellow swimming caps, three were wearing red swimming caps and the rest were wearing black swimming caps*. Ask pupils: *How many children were wearing black swimming caps?* (Pupils should extend their recording to 8 – 2 – 3 = ☐.) Repeat the story to give pupils time to complete their recording.
- Challenge pupils to try a different method to find the missing part.
- Ask pupils to complete Question 4 in the Practice Book.

Same-day intervention

- Give pupils a ten-frame and some counters or small counting objects.
- Explain that you are going to tell them a story and you would like them to model it as you are going along so that they can quickly find the answer.
- Begin with: *There were ten balloons blown up ready for the party*. Check that pupils recognise that they can represent the balloons using counters on their ten-frame.
- Continue the story with: *A gust of wind blew three balloons away*. Ask pupils how they can represent this on their ten-frame. Agree that they can remove three counters to represent the three balloons. Encourage pupils to do this from the bottom right-hand corner and along the bottom row so that they can see and recognise the layout of 5 and 2 more, 7 counters left. Record this together as 10 – 3, agreeing to stop there until you have told the rest of the story.
- Continue the story with: *Two balloons suddenly popped. How many balloons were left for the party?* Ask pupils how they can represent this on their ten-frame and agree that they can remove two more counters, again working along the bottom row. With the top row of counters complete, pupils may be able to quickly tell you that there are five balloons left for the party.
- Record this as 10 – 3 – 2 = 5. Ask pupils to say this number sentence with you as they point to the relevant counters.
- Ask pupils to return their counters to the ten-frame and tell them another story. Ask pupils to record the story in a number sentence as they move their counters. Support as necessary.

Chapter 2 Addition and subtraction within 10

Unit 2.14 Practice Book 1A, pages 70–71

Same-day enrichment

- Give pupils two sets of 0–4 digit cards and a 5.
- Ask pupils to work in pairs. They should shuffle the cards. The first person turns over one card at a time, telling a subtraction story from 9 as they do so, but not showing the cards to their partner. Their partner keeps track of the problem using cubes or a number line, telling their partner the difference. The storyteller then reveals the two cards and they check the calculation together before swapping roles.
- Remind pupils of the list of topics they drew up for their addition stories. Add any other topics that pupils used.
- Make sure that pupils have the opportunity to tell at least two stories each. This will help to consolidate understanding of the structure of written problems.

Challenge and extension question

Question 5

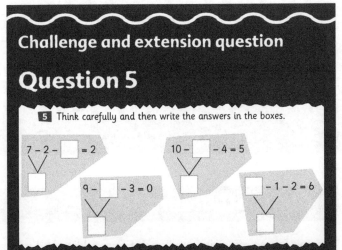

5 Think carefully and then write the answers in the boxes.

$7 - 2 - \square = 2$

$10 - \square - 4 = 5$

$9 - \square - 3 = 0$

$\square - 1 - 2 = 6$

This question encourages pupils to explore mixed additions and subtractions in the same number sentence. Some pairs of numbers are linked to support calculating. The operations signs are provided throughout but pupils will need to have recognised the inverse relationship between addition and subtraction as they will often need to work backwards, demonstrating both fluency and reasoning, to find the value of a missing number.

Chapter 2 Addition and subtraction within 10

Unit 2.15
Mixed addition and subtraction

Conceptual context

This unit focuses on giving pupils practice at adding and subtracting three numbers within 10, building on Units 2.13 and 2.14. In this unit, addition and subtraction are mixed within the same calculation.

All previous approaches are revisited and combined so that pupils can link and apply them to solve a number sentence with two different operations. Inequalities are also revisited and used with the mixed operation number sentences.

Having worked through Chapter 2, pupils will have been introduced to a variety of representations to support addition and subtraction within 10. The Chapter 2 test will check that all these skills are embedded before pupils extend their skills to working with numbers from 11 to 20 in Chapter 3.

Learning pupils will have achieved at the end of the unit

- Pupils will have revisited adding and subtracting three numbers within 10 using a variety of representations (Q1, Q2, Q3, Q4)
- Pupils will have explored and consolidated adding and subtracting within the same calculation (Q1, Q2, Q3, Q4)
- Pupils will have recognised that they can record addition and subtraction within the same number sentence (Q1, Q2)
- Pupils will have explored and practised recording a mixed calculation on the same number line (Q2)
- Pupils will have consolidated comparing quantities, recording the comparisons with <, > and = (Q3)
- Pupils will have revisited using <, > and = to express numerical relationships (Q3)
- Pupils will have explored a further representation to support addition and subtraction, the bead string (Q4)

Resources

sheets of paper; paper plates; ten-frames; 0–9 digit cards; 0–10 number lines; bead strings; counters; interlocking cubes; small counting objects

Vocabulary

row, add, addition, subtract, subtraction, equals, equivalent, inequality

Chapter 2 Addition and subtraction within 10

Unit 2.15 Practice Book 1A, pages 72–74

Question 1

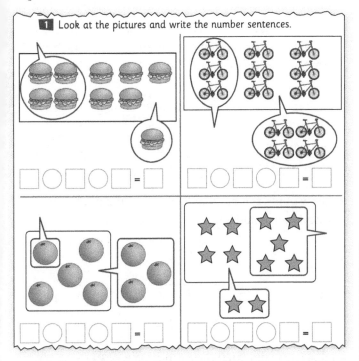

What learning will pupils have achieved at the conclusion of Question 1?

- Pupils will have revisited adding and subtracting three numbers within 10 using a variety of representations.
- Pupils will have explored adding and subtracting within the same calculation.
- Pupils will have discovered that they can record addition and subtraction within the same number sentence.

Activities for whole-class instruction

- Give each pupil a sheet of paper and ten objects. Explain that the sheet of paper is going to be their working area. Ask them to count eight objects onto their paper, arranging them in a way that makes it easy to see how many there are. A row of 5 and a row of 3 or in pairs would be helpful. Explain that you are going to work through some additions and subtractions together, recording the matching number sentences. Explain that sometimes you will add then subtract, other times you might subtract then add.
- Ask pupils to begin their number sentence by recording how many they have, 8. Ask them to subtract 3, showing this on the paper and in their partly completed number sentence. Remind pupils to remove objects in such a way that the arrangement on the paper supports recognition of how many without counting. Pupils should record 8 − 3.

- Now ask pupils to add two more objects to their paper, positioning them in a way that helps them see how many they have altogether. Pupils should record 8 − 3 + 2 = 7. Read the number sentence together, pointing to relevant counters.
- Work through another example, beginning with a smaller number of counters and adding, then subtracting.
- Ask pupils to look at the first illustration in Question 1. Check that they recognise when an amount is being added, the arrow points onto the rectangle and when it is being subtracted, the arrow points away from the rectangle.

Same-day intervention

- Give pupils a ten-frame and ten objects. Count the objects onto the ten-frame to remind pupils that the frame has ten spaces and to check that they have ten objects.
- Ask pupils to clear their ten-frame. Explain that you are going to do some additions and subtractions together. Ask pupils to put three objects on their frame. Remind them to start at the top left so that they create a familiar picture that they can recognise on the frame. Ask pupils to record how many they have on their frame, 3.
- Now ask pupils to add two more objects to their ten-frame. Encourage them to continue placing them on the top row until it is full before starting on the bottom row. Ask pupils to record what they have done so far, but not to record the sum as they have not finished yet. Pupils should record 3 + 2. Ask pupils how many objects they have on their frame. Remind them that the top row is full, so they may be able to tell you without counting. Confirm that there are five, because the ten-frame is two rows of 5 and they have filled one of them.
- Now ask pupils to subtract four objects from their ten-frame. Ask pupils to complete the rest of their number sentence to show what they did: 3 + 2 − 4 = 1.
- Ask pupils to remove all the counters from their ten-frame and then act out what they did with counters as you say the number sentence: 3 + 2 − 4 = 1.
- Repeat, beginning with a larger number and first subtracting, then adding.

Chapter 2 Addition and subtraction within 10

Unit 2.15 Practice Book 1A, pages 72–74

Same-day enrichment

- Give each pair of pupils a set of 0–9 digit cards and a large version of the number sentence framework, ☐ + ☐ − ☐ = ☐, big enough for the digit cards to fit in the boxes.
- Challenge pupils to make as many different correct number sentences as they can.
- Remind them that they only have one of each number to 9, so they need to think carefully about what they can and cannot do.

Question 2

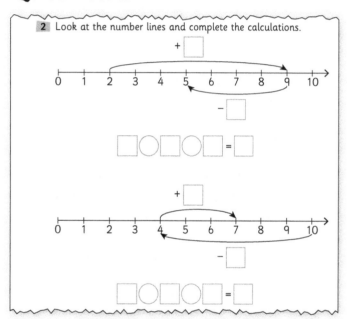

What learning will pupils have achieved at the conclusion of Question 2?

- Pupils will have explored and practised recording a mixed calculation on the same number line.
- Pupils will have continued to record addition and subtraction within the same number sentence.

Activities for whole-class instruction

- Give pupils a number line and show them the number sentence 4 + 5 − 7 = ☐.
- Explain that, if both actions were additions, they could simply include another jump along the number line. If both actions were subtraction, they could also use a second jump, but with one addition and one subtraction, it could be rather confusing.
- To avoid any confusion, pupils could carry out the addition as normal, but then jump back below the line.
- Work through the calculation together, first completing the addition by making a jump of 5 from 4 to 9. Then subtract 7 from 9 using a jump back of 7 from 9 below the line.

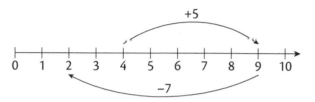

- Show pupils this number line and ask them what the calculation could be.

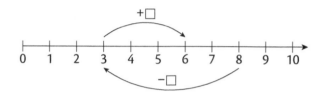

- Pupils may think that the calculation was 3 + 3 − 5, but the correct solution of 1 is not what is shown on this number line. Draw pupils' attention to the fact that the subtraction arrow begins at 8, not 6. The subtraction must be the first operation: 8 − 5 + 3 = 6.
- Explain that when pupils see this layout, they need to look carefully to identify where the calculation begins and ends as well as the order of the operations or the number sentence they record will not match the number line. The direction of the arrows is important and will help.
- Pupils should now be ready to work through Question 2.

Same-day intervention

- Remind pupils that they have already carried out some calculations with both an addition and a subtraction using a ten-frame. Explain that they are going to explore carrying out a similar calculation on a number line.
- Show pupils the calculation 3 + 5 − 4.
- Ask pupils to begin by getting three cubes and joining them together to make a tower. Count along the number line to 3 and ask pupils to underline 3 to remind them that they have three cubes.
- Now ask them to make a tower of five cubes in a different colour. Refer to the number sentence and point out that it begins with 3 + 5. Ask pupils to join the 5 tower to the 3 tower and count how many cubes they have.

Chapter 2 Addition and subtraction within 10

Unit 2.15 Practice Book 1A, pages 72–74

- Confirm that they have 8. Remind pupils how to record this as a jump of 5 from 3 to 8 on their number line.

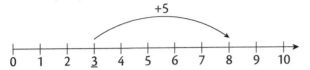

- Return to the number sentence, 3 + 5 – 4. Explain that pupils have recorded the first part of the calculation and now they are ready to move on to the second part. This says subtract 4. Ask pupils to remove four of the five cubes and to count how many they have left, 4.

- Show pupils that this can be recorded on the number line as a jump below the line, so that the two actions do not become muddled.

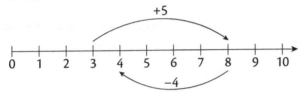

- Show pupils this number line and ask them what the calculation could be.

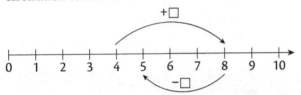

- Agree that the first jump is + 4, from 4 to 8 and the second jump is – 3, from 8 to 5. So the calculation is 4 + 4 – 3 = 5. Ask pupils to check this by making a tower of four cubes, adding a second tower of four cubes to make a tower of 8, then to remove three of those cubes to leave 5. The number sentence is correct.

- Work through the first number line of Question 2 together. Give pupils time to look at the second number line. It may be necessary to support pupils by underlining 10 to help them identify that the number sentence starts at 10 and the first action is subtraction.

Same-day enrichment

- Ask pupils to draw a double function machine to represent each of the calculations in Question 2. Ask: *Which numbers are between the machines? What do you notice about those numbers on the number lines in the Practice Book question?*

- Ask pupils to make up a word problem to match the two problems they have represented with number lines and function machines.

Question 3

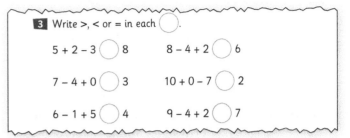

What learning will pupils have achieved at the conclusion of Question 3?

- Pupils will have consolidated comparing quantities, recording the comparisons with <, > and =.
- Pupils will have revisited using <, > and = to express numerical relationships.

Activities for whole-class instruction

- Give pupils access to a number line, ten-frame and cubes.
- Show pupils the partly completed number sentence 8 – 3 + 5 and ask them to find a solution in any way they choose. Confirm that the total is 10, but write the number statement 8 – 3 + 5 ◯ 7. Ask pupils what they could write in the circular box to make the expression true. Confirm that they cannot write =, because both sides of the equals sign are not equivalent in value. Confirm that they could use >, is greater than, because 10 is greater than 7.
- Work through another example before asking pupils to complete Question 3 in the Practice Book.

Same-day intervention

- Show pupils a calculation with both addition and subtraction, for example 3 + 5 – 4 = ☐. Read together as *3 add 5 subtract 4 = something*.
- Ask pupils to begin by making a tower of three cubes. Ask pupils what the calculation asks them to do next. Confirm that they need to add 5. Ask pupils to make a tower of five cubes in a different colour to the first one. After joining the two towers, ask pupils to count their cubes. Confirm that they have 8, because 3 + 5 = 8.
- Return to the calculation. Agree that the last part of the calculation asks pupils to subtract 4 from what they have. Ask pupils to remove four cubes from the five they added and to check how many cubes they have left.

Chapter 2 Addition and subtraction within 10

Unit 2.15 Practice Book 1A, pages 72–74

- Agree there are four cubes left, so the completed number sentence is 3 + 5 − 4 = 4.
- Rewrite the number sentence as: 3 + 5 − 4 ◯ 6.
- Remind pupils that square boxes are used for missing numbers and circular boxes are used for signs such as +, −, =, < and >. Pupils already know that 3 + 5 − 4 = 4, so the equals sign cannot be the missing symbol. Remind pupils that < stands for 'is less than' and > stands for 'is greater than'. Read the number sentence together using 'is less than' and 'is greater than', asking pupils which is correct.
- Since 3 + 5 − 4 = 4, 3 + 5 − 4 is less than 6, so the completed number sentence must be 3 + 5 − 4 < 6.
- Work through another example together.
- Give pupils cubes to use to help them complete Question 3. Encourage them to find a solution. They then compare that with the number provided to help them identify the correct sign to make the number sentence correct.

Question 4

4 Complete the following calculations.

2 + 4 + 3 = ☐   6 − 3 + 4 = ☐

8 − 4 + 2 = ☐   6 + 4 + 0 = ☐

2 + 2 − 1 = ☐   1 + 6 − 5 = ☐

8 − 7 − 0 = ☐   10 − 3 + 2 = ☐

2 + 7 − 3 = ☐   2 + 5 + 1 = ☐

4 − 2 + 8 = ☐   6 − 4 + 5 = ☐

What learning will pupils have achieved at the conclusion of Question 4?

- Pupils will have explored a further representation to support addition and subtraction, the bead string.
- Pupils will have revisited adding and subtracting three numbers within 10 using a variety of representations.
- Pupils will have consolidated adding and subtracting within the same calculation.

Activities for whole-class instruction

- Give pupils bead strings with ten beads, ideally with five beads in one colour and five in another. If your bead strings are longer, tie the rest of the bead string off so that pupils only work with ten beads. As pupils work with greater numbers, the string can be untied.
- Count along the ten beads together, forwards and back.
- Explain that the bead string is similar to a number track. It is like a tower of cubes on a string.
- Display a calculation such as 6 − 3 + 4. Ask pupils to start with all the beads on their right, then move 6 to the left to the very end of the string to represent the 6 at the start of the calculation. From that 6, ask pupils to subtract 3. Pupils do this by moving three beads to the right to join the other beads there. The final step, + 4, is achieved by sliding four beads back to the left to join those there. All steps have been carried out, so the number of beads on the left is the solution to the calculation.
- If the bead strings have five beads of one colour and five of another, focus on 6 being five beads of one colour and 1 more, and on using 5 and some more to count how many beads in the solution.

Same-day enrichment

- Create a grid as follows. Print out and cut along the grid lines.

10	− 4	+ 3	> 7
9	− 7	+ 0	= 2
8	− 3	+ 1	< 9
7	− 5	+ 5	> 5
6	+ 2	− 6	= 2
5	+ 4	− 2	< 10

- Give pairs of pupils a copy each. Explain that there are six number statements. Two use the equals sign, two use 'is less than' < and two use 'is greater than' >. Challenge pupils to find six completed number statements.
- Ask pupils to compare results. Are their solutions all the same? Challenge pupils to record their solution and find a different one.
- If pupils need some help to get started, suggest they complete two number sentences that equal 2 first of all, using some of the larger numbers, although they may need to adjust part of these number sentences.

Chapter 2 Addition and subtraction within 10

Unit 2.15 Practice Book 1A, pages 72–74

- Pupils may also recognise that they have seven beads on the left because there are three on the right and they know that 7 + 3 = 10.
- Complete two more mixed calculations, reading the calculation together and carrying out each addition and subtraction as it is said. Repeat to check.
- Pupils should use the bead string to help them complete Question 4.

Same-day intervention

- Remind pupils that they have already seen calculations with two additions, two subtractions or one of each. They have also explored how to solve these using a ten-frame and counters, cubes, a number line and a bead string.
- Show pupils a calculation such as 8 − 5 + 2 = ☐. Read together as *8 subtract 5 add 2 equals something*. Give pupils a ten-frame and some counters and ask them to explain how to use this to solve the calculation.
- Agree that they need to start with eight counters, because this is how the calculation starts. Give pupils time to place eight counters on their ten-frame, using the top row and part of the bottom row, beginning from the left. Agree that the next part of the calculation is subtract 5. Give pupils time to remove five of their counters, all from the bottom row and two from the top right. Ask pupils how many counters they have left and what they still need to do to complete the calculation.
- Agree that pupils have three counters left. Show them how to use the number sentence to record what they have done so far. Since they have carried out the 8 − 5 part of the calculation, it is useful to lightly record 3 above the + sign, to remind them that they have three counters so far before they add a further two.

$$\overset{3}{8 - 5} + 2 = \square.$$

- Ask pupils to complete the calculation and the matching number sentence: 8 − 5 + 2 = 5.
- Explain that whatever method they use, they might find it useful to make a temporary record of the result of the first part of the calculation to help them keep track of where they are in the calculation.
- Support pupils to use any method to complete the calculations in Question 4.

Same-day enrichment

- Show pupils the two patterns of calculations:

  1 + 2 − 3 = 0   10 − 9 + 8 = 9
  2 + 3 − 4 =   9 − 8 + 7 =
  3 + 4 …   8 − 7 …

- Ask pupils to continue the calculations, exploring the patterns as they solve their calculations.
- Can pupils explain what is happening? In the first pattern, all three numbers are increased by 1 in each new calculation, but 2 are added and 1 is subtracted, so the total increases by 1. In the second pattern, all three numbers are decreased by 1 but 2 are subtracted and 1 is added, so the total decreases by 1. Pupils may recognise the patterns but find it difficult to give reasons for them.

Challenge and extension question

Question 5

5 Think carefully and then write the answers in the boxes.

4 + 4 − ☐ = 7   ☐ − 2 + 1 = 8
       ☐   ☐

3 + ☐ − 2 = 6   ☐ − 1 + 2 = 2
       ☐   ☐

This question encourages pupils to explore mixed additions and subtractions in the same number sentence. Some pairs of numbers are linked to support calculating. Although pupils might expect to calculate with the first pair of numbers initially, often one of those numbers is missing. The operations signs are provided throughout but pupils will need to have recognised the inverse relationship between addition and subtraction as they will often need to work backwards, demonstrating both fluency and reasoning, to find the value of a missing number.

Chapter 2 Addition and subtraction within 10

Unit 2.15 Practice Book 1A, pages 72–74

Chapter 2 test (Practice Book 1A, pages 75–77)

Test question number	Relevant unit	Relevant questions within unit
1	2.3	3
	2.4	3
	2.5	3
	2.6	2
	2.7	4
	2.8	3
	2.9	3
	2.11	3
2	2.2	4
	2.12	4
3 (a), (b)	2.11	2
3 (c)	2.15	2
4	2.13	3
	2.14	3
	2.15	4
5	2.4	1
	2.5	1
	2.6	1
	2.7	2
	2.8	1
	2.10	1
	2.13	1
	2.14	1

Chapter 3
Numbers up to 20 and their addition and subtraction

Chapter overview

Area of mathematics	National Curriculum statutory requirements for Key Stage 1	Shanghai Maths Project reference
Number – number and place value	Year 1 Programme of study: Pupils should be taught to: ■ given a number, identify one more and one less	Year 1, Unit 3.3
	■ identify and represent numbers using objects and pictorial representations including the number line, and use the language of: equal to, more than, less than (fewer), most, least	Year 1, Units 3.1, 3.2, 3.3, 3.4, 3.5, 3.6, 3.7, 3.8, 3.9, 3.10, 3.11
	■ read and write numbers from 1 to 20 in numerals and words.	Year 1, Units 3.1, 3.2, 3.3, 3.4, 3.5, 3.6, 3.7, 3.8, 3.9, 3.10, 3.11, 3.12
	Year 2 Programme of study: Pupils should be taught to: ■ compare and order numbers from 0 up to 100; use <, > and = signs	Year 2, Units 3.4, 3.6, 3.8
	■ use place value and number facts to solve problems.	Year 2, Units 3.2, 3.4, 3.6, 3.8, 3.9, 3.10, 3.11, 3.12

Area of mathematics	National Curriculum statutory requirements for Key Stage 1	Shanghai Maths Project reference
Addition and Subtraction	Year 1 Programme of study: Pupils should be taught to:	
	■ read, write and interpret mathematical statements involving addition (+), subtraction (−) and equals (=) signs	Year 1, Units 3.1, 3.2, 3.3, 3.4, 3.5, 3.6, 3.7, 3.8, 3.9, 3.10, 3.11
	■ represent and use number bonds and related subtraction facts within 20	Year 1, Units 3.4, 3.6, 3.8, 3.9, 3.10, 3.11, 3.12
	■ add and subtract one-digit and two-digit numbers to 20, including zero	Year 1, Units 3.1, 3.2, 3.3, 3.4, 3.5, 3.6, 3.7, 3.8, 3.9, 3.10, 3.11, 3.12
	■ solve one-step problems that involve addition and subtraction, using concrete objects and pictorial representations, and missing number problems such as $7 = \square - 9$.	Year 1, Units 3.2, 3.4, 3.6, 3.8, 3.9, 3.10, 3.11, 3.12
	Year 2 Programme of study: Pupils should be taught to	
	■ recall and use addition and subtraction facts to 20 fluently, and derive and use related facts up to 100	Year 2, Units 3.4, 3.6, 3.8, 3.9, 3.11, 3.12
	■ show that addition of two numbers can be done in any order (commutative) and subtraction of one number from another cannot	Year 2, Units 3.6, 3.7, 3.8, 3.9, 3.11
	■ recognise and use the inverse relationship between addition and subtraction and use this to check calculations and missing number problems.	Year 2, Units 3.11, 3.12

Chapter 3 Numbers up to 20 and their addition and subtraction

Unit 3.1
Numbers 11–20

Conceptual context

This is the first in a series of units on using numbers between 10 and 20 and builds on pupils' experience of working with numbers up to 10. The focus initially is on recognising that each of the numbers between 11 and 20 comprises 1 'ten' and 'some more' (ones). This helps to develop and consolidate their familiarity with tens as a visual representation, and use the representations to help strengthen their ability to subitise and develop instant recall of number facts involving 10 and some more.

Although it is likely that pupils will have much experience with oral counting of numbers greater than 10, it is important that they develop their understanding of the tens and ones within each number through extensive first-hand experience of these quantities. This is necessary as it underpins later work on place value and partitioning into tens and ones, and helps to ensure that pupils' understanding of the numerals relates to quantities they can visualise, not just abstract symbolic notation. It also helps to develop their repertoire of known and quickly derived facts relating to numbers between 10 and 20. Consequently, it is vital that pupils understand that '14' represents 1 ten and 4 ones, and not a 1 and a 4. Such understanding is developed through extensive first-hand experience of counting and matching a range of representations to appropriate numerals alongside constant reinforcement of appropriate mathematical terms.

Learning pupils will have achieved at the end of the unit

- Pupils will have extended their counting skills and application of the counting principles to numbers beyond 10 (Q1)
- Pupils will be able to write the numeral that matches a quantity (Q1)
- Pupils will have developed their ability to use the arrangement of objects to help them to subitise quantities (Q1)
- Pupils will have developed their experience with number lines beyond 10 (Q2)
- Pupils will have applied their knowledge of < and > symbols to numbers between 10 and 20 (Q2)
- Pupils will be able to recognise, describe and continue patterns with numbers between 0 and 20 (Q3)
- Pupils will have consolidated their understanding that numbers can be made up of some tens and some ones (Q3)

Resources

variety of everyday objects to count (e.g. fruit, shells, pencils, toy cars, stickers, marbles, rubber ducks in a water tray); mathematics resources such as counting cubes; early base 10 resources such as coloured 1–10 flat pieces and matching pegs; pictures of quantities from 1 to 20 and numeral cards on a number line (or washing line); arrow cards; mini whiteboards

Vocabulary

ten, 10 and some more, 11, 12, 13, 14, 15, 16, 17, 18, 19, 20, eleven, twelve, thirteen, fourteen, fifteen, sixteen, seventeen, eighteen, nineteen, twenty, odd number, even number, greater than, less than

Subitising

To be able to recognise how many objects are in a set without the need to count them. Typically, we can subitise quantities to around 7, although this can vary depending on how the objects are presented. Dice layout and arrays help to make subitising easier.

Chapter 3 Numbers up to 20 and their addition and subtraction Unit 3.1 Practice Book 1A, pages 78–79

Question 1

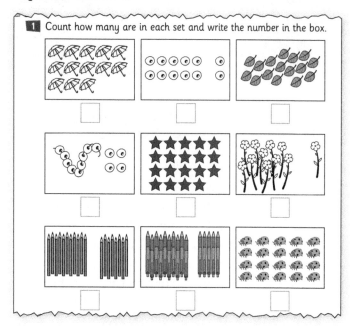

What learning will pupils have achieved at the conclusion of Question 1?

- Pupils will have extended their counting skills to beyond 10, recognising that a number can be 10 and some more.
- Pupils will be able to write the numeral that matches a quantity.
- Pupils will have developed their ability to use the arrangement of objects to help them to subitise quantities.

Activities for whole-class instruction

- In order to rehearse the counting principles (one-to-one correspondence, stable order principle, cardinality, order irrelevance – described more fully in Unit 1.4), the class count together some fruit (or other objects) out of a bowl. Count out ten pieces of fruit and arrange them into two rows of 5. Remind the class that we can see that there are ten pieces of fruit because there are two groups of 5 (in order to help develop subitising). Once ten pieces of fruit have been counted, continue to count saying: *10 and 1 more, 10 and 2 more, 10 and 3 more, 10 and 4 more* and so on, up to 19.
- Give pairs of pupils a selection of real-world objects to count. First, they should count out ten of the objects then rehearse saying: *10 and 1 more, 10 and 2 more, 10 and 3 more, 10 and 4 more* and so on, up to 19.
- With the whole class, repeat the previous counting task, this time saying: *1 ten and 1 more is equal to 11, 1 ten and 2 more is called 12* and so on.

- With the whole class, repeat the counting out task, this time adding numerals on arrow cards for each group. Count out the first group to 10 then place a 10 arrow card to indicate 10. Continue to count out one move piece of fruit. Say: *10 and 1 more* and lay out the arrow card for 1, then when you say: *is called 11*, overlap the arrow cards so that pupils can see that the number 11 comprises 1 ten and a 1. Repeat for other teens numbers.

 ... pupils who are not applying the counting principles correctly: not giving each item one number label (one-to-one principle); not remembering the number names in the correct order (stable order principle); not recognising that the last number said in the count is the number in the set (cardinal principle); not being able to transfer their counting skills to different groups of objects (the abstraction principle); finding it difficult to count objects that are arranged in groups of circles (order irrelevance principle).

(All say…) *10 and xxxx more is equal to xxxxteen.*

Same-day intervention

- Provide additional opportunities for pupils to count physical objects in order to consolidate their use of the counting principles. This should include counting all of the objects in a set and counting out a particular quantity of objects from a larger set.
- Provide additional opportunities to count objects to match written numerals.
- Provide additional opportunities to match arrow cards to quantities of objects and to rehearse the language of *10 and xxxx more is equal to xxxxteen.*

Same-day enrichment

- Ask questions varying the position of the knowns and unknowns in questions expressed symbolically and in words, for example:

  10 + ? = 15

  ? + 10 = 15

  10 + 5 = ?

  15 = 5 + ?

  15 = ? + 10

  – If I have 15 oranges, how many more than 10 is that?
  – What number is 6 more than 10?

Chapter 3 Numbers up to 20 and their addition and subtraction Unit 3.1 Practice Book 1A, pages 78–79

Question 2

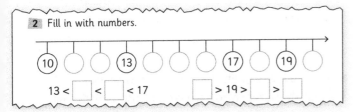

What learning will pupils have achieved at the conclusion of Question 2?

- Pupils will have developed their experience with and use of number lines beyond 10.
- Pupils will have applied their knowledge of < and > symbols to numbers between 10 and 20.

Activities for whole-class instruction

- Washing line number line activities. This question focuses on the symbolic representation of numbers between 10 and 20 but pupils should also continue to develop their experience of associating numerals with quantities.

- Using pictures of quantities, rehearse the oral counting developed for Question 1 (*10 and 1 more is 11, 10 and 2 more is 12* and so on). Ideally these pictures of quantities should aid subitising by arranging the images in familiar arrangements (lines of five, for example). Place these pictures on a 'washing line' number line. Now match the numeral to each quantity so that pupils can associate the image with what each number is called.

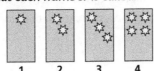

- Once the number line from 1 to 20 is completed, ask pupils questions such as:

  Who can show me the number that is 10 and 6 more? Who can tell me the quantity shown here? Who can find me the picture that shows 17? If I have 8 more than 10, what is that number called? Who can show me a picture of 18? and then: *Who can tell me a number which is greater than 13? Who can tell me a number which is less than 18? Who can tell me the number which is equal to 10 and 3 more?*

- Model writing these responses using < , > and = symbols to help re-familiarise pupils with them. Once pupils are using the symbols accurately, they can complete Question 2 in the Practice Book to consolidate and record what they have learned.

 ... pupils who are not yet fluent with teens numbers, for example write 41 instead of 14.

Same-day intervention

- If pupils cannot use numerals reliably then continue to provide additional experience with counting out objects into 10 and some more. Match arrow cards to the quantities, combine and say: *10 and 4 more is called 14.*

Same-day enrichment

- Extend pupils' understanding of the relative size and position of numbers by asking them to find more than one number that is, for example >12 and <16.

Question 3

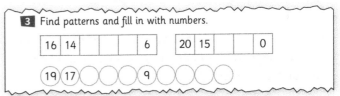

What learning will pupils have achieved at the conclusion of Question 3?

- Pupils will be able to recognise, describe and continue patterns with numbers between 0 and 20.
- Pupils will have consolidated their understanding that numbers can be made up of some tens and some ones.

Chapter 3 Numbers up to 20 and their addition and subtraction

Unit 3.1 Practice Book 1A, pages 78–79

Activities for whole-class instruction

- Pupils complete a whole-class counting task, using a range of different objects. Again, count out 10, position the '10' arrow card, then some more and position the appropriate arrow card beneath the set. Continue with numbers up to and including 20.

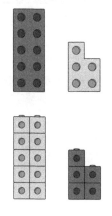

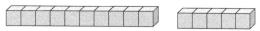

Repeat the task, this time positioning counting materials such as coloured tiles, connecting cubes or base 10 blocks to represent the ten.

All say... 10 and xxxx more is xxxxteen, up to 2 tens is equal to 20.

(i) Teaching with variation involves carefully varying the contexts and resources used when developing pupils' understanding of a concept. According to Gu (1999), it is an important teaching method through which students can definitely master concepts. It intends to illustrate the essential features by demonstrating different forms of visual materials and instances or highlight the essence of a concept by varying the nonessential features. It aims at understanding the essence of object and forming a scientific concept by putting away nonessential features (Gu, M., (1999) Education directory. Shanghai: Shanghai Education Press).

- Write teens numbers on the board (10, 12, 14, 16, 18), (11, 13, 15, 17, 19), (10, 15). For each number, say its name and how many tens and ones make up the number (if appropriate, match again to a representation of the quantity).

- Ask pupils to say what they see and what they notice about the numbers in each group. Ensure that pupils answer in sentences; in the first group each number is 2 more than the number before and in the second group each number is 2 more than the number before.

- Use the features of the various manipulatives to emphasise the structure of the numbers and support pupils' explanations.

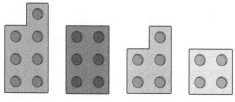

- Reverse the order of the numbers to count backwards from 20 and repeat. Pupils should respond: *Each number is 2 less than the number before.* Ask if the set is still made up of even numbers and set two still made up of odd numbers. Pupils should respond in full sentences.

All say... I know that 12 is an even number because it can be laid out in pairs. I know that 11 is an odd number because it cannot be laid out in pairs.

 ... pupils reciting the numbers in the patterns in order but not able to associate them with a visual representation or describe them in terms of some more than, or some less than, the number that preceded it.

Chapter 3 Numbers up to 20 and their addition and subtraction   Unit 3.1 Practice Book 1A, pages 78–79

Same-day intervention

- Continue to practise counting out objects in twos (from 2 to 20) to help pupils become more fluent with counting in even numbers and to associate the number with a visual representation. Count out objects in ones, positioning them to make pairs. With each additional object, highlight that this number is now 1 more than a multiple of 2 and is an odd number.

Same-day enrichment

- Ask pupils to sort the following into one of two groups: 'odd' or 'even'.

  1 ten and 4 more

  15

  12

  2 tens and 3 more

  35

  32

- Pupils should identify numbers around the school that are odd and even (classroom numbers, door numbers). They should look at larger numbers such as those on number plates to lead to a generalisation that the number of ones determines whether a number is even or odd.

Challenge and extension question

Question 4

4 (a) The kangaroo hops on odd numbers (11, 13, …). Help the kangaroo draw a green line to link the numbers from the least to the greatest.

The frog hops on even numbers (10, 12, …). Help the frog draw a red line to link the numbers from the greatest to the least.

15 20
14 16 17 18
12 13 11 10 19

(b) Write the numbers from (a).

Odd numbers: ☐

Even numbers: ☐

This question helps to develop fluency in recognising odd and even numbers by presenting them in a jumbled order.

On their whiteboards, pupils respond to the following questions.

Write down the number that is:

- 1 ten and 3 more
- 5 more than 10
- an even number less than 20
- an odd number between 10 and 20
- an even number greater than 10 and less than 15
- an odd number that is 1 less than 20
- I am thinking of a number, it is an even number that is greater than 16 and less than 20, what is my number?

Chapter 3 Numbers up to 20 and their addition and subtraction

Unit 3.2
Tens and ones

Conceptual context

This lesson focuses on developing pupils' understanding of tens and ones and being able to partition two-digit numbers less than 20 into 10 and some more. Pupils' experience of doing this is broadened to include using physical resources, pictures and symbols and develops their understanding of +, − and = symbols, through their use with pictures and numerals. Their fluency with these numbers is applied to solving simple addition and subtraction questions.

Learning pupils will have achieved at the end of the unit

- Pupils will have developed greater fluency in describing numbers between 10 and 20 (10 and some more) (Q1, Q2)
- Pupils will have developed greater fluency in matching a numeral to a quantity (Q1, Q2, Q3, Q4)
- Pupils will have consolidated their understanding of place value and will recognise that two-digit numbers comprise tens and ones (Q1, Q2)
- Pupils will have begun to use the word 'partition' when explaining how numbers can be separated into tens and ones (Q1, Q2, Q3, Q4)
- Pupils will have become familiar with bar-model representations of numbers (Q2)
- Pupils will have begun to count on rather than count all to find a total (Q3)
- Pupils will have developed their fluency in partitioning numbers into 10 and some more and will apply this to addition and subtraction questions (Q2, Q3, Q4)

Resources

variety of everyday objects to count (e.g. shells, pencils, toy cars, stickers, marbles, pictures of quantities from 1 to 20); counters; base 10 blocks; numeral cards on a number line (or washing line); mini whiteboards; coins (1p, 2p, 5p, 10p); arrow cards

Vocabulary

ten, 10 and some more, tens and ones, 11, 12, 13, 14, 15, 16, 17, 18, 19, 20, eleven, twelve, thirteen, fourteen, fifteen, sixteen, seventeen, eighteen, nineteen, twenty, odd number, even number, subtract, add, is equal to

Chapter 3 Numbers up to 20 and their addition and subtraction   Unit 3.2   Practice Book 1A, pages 80–81

Question 1

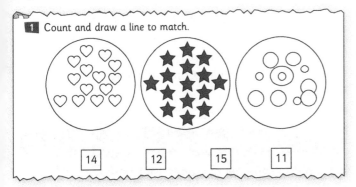

What learning will pupils have achieved at the conclusion of Question 1?

- Pupils will have developed greater fluency in describing numbers between 10 and 20 (10 and some more).
- Pupils will have developed greater fluency in matching a numeral to a quantity.
- Pupils will have consolidated their understanding of place value and will recognise that two-digit numbers comprise tens and ones.
- Pupils will have begun to use the word 'partition' when explaining how numbers can be separated into tens and ones.
- Pupils will have become familiar with bar-model representations of numbers.

Activities for whole-class instruction

- Introduce a whole-class counting task. Ask pupils to take a handful of shells (or other small objects) and to estimate how many there are. Ask pupils to suggest a good way of counting how many there are. Establish that we can make up a group of 10 and then see how many more than 10 there are. Repeat for quantities from 11 to 20. Explain and demonstrate that the quantity (for example 16) can be partitioned into 1 ten and 6 ones.
- Ask pupils to say the name for each of the quantities made and use arrow cards to illustrate how each number would be written.
- Pupils should then complete Question 1 in the Practice Book.

 10 and xxxx more is equal to xxxxteen.

Same-day intervention

- Introduce additional opportunities for pupils to count physical objects to consolidate their '10 and some more' fluency and match physical quantities to the number names and the appropriate numerals.

Same-day enrichment

- Look at teens numbers in real-life contexts. Using photos of windows on buildings, tins on a supermarket shelf, groups of people, flocks of birds, dishes of sweets, cars in a car park, fish in a tank and so on, pupils should circle a group (or groups) of 10 and count the number of ones. They write the correct numeral.

Look out for … pupils who are 'mis-naming' numbers (for example 41 instead of 14). Ask them: *Is this forty-one or fourteen? Does it have 4 tens (point to the 4 in 41) or 1 ten (point to the 1 in 14)?*

Question 2

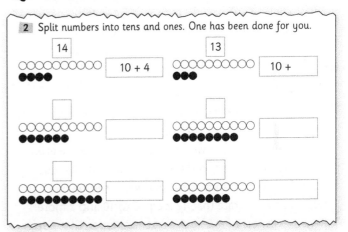

What learning will pupils have achieved at the conclusion of Question 2?

- Pupils will have developed greater fluency in describing numbers between 10 and 20 (10 and some more).
- Pupils will have begun to use the word 'partition' when explaining how numbers can be separated into tens and ones.
- Pupils will have developed greater fluency in matching a numeral to a quantity.
- Pupils will have become familiar with bar-model representations of numbers.

Chapter 3 Numbers up to 20 and their addition and subtraction   Unit 3.2   Practice Book 1A, pages 80–81

Activities for whole-class instruction

- It is important that pupils develop a secure understanding of 10 and some more and can recognise that a teens number is one group of 10 and a group of ones without always needing to see the group of 10.

- This familiarity with the idea that 10 ones is equal to 1 ten is a significant one and builds on their understanding that objects can have different appearances but represent the same value (for example two 5p coins and one 10p coin). It encourages counting on rather than counting all and also underpins the concept of exchange, which pupils will be introduced to when they are introduced to more formal written strategies for addition and subtraction. To help pupils develop their understanding of 10 ones being equal to 1 ten, this task focuses on exchanging 10 ones for 1 ten as part of identifying a tens and ones number.

- Use base 10 blocks (the ones and the tens sticks) to show pupils that 10 ones can also be conserved as 1 ten. Pupils should line up 10 ones against a 10 stick so they see for themselves that they are the same length and can be imagined as stuck together. Ask pupils to close their eyes and imagine a group of 10 ones floating together to make a 10 stick. Then they should imagine the 10 exploding into 10 ones.

  All say... 10 ones is the same as 1 ten.

- Lay out a quantity of counters, such as 12. Clearly identify that within this group, there is one group of 10 and 2 more counters by drawing a rectangle around the group of 10 and another around the 2. Explain that 12 is equal to 10 and 2 more. Draw another bar below the 10 and 2 and write 12 to illustrate that 12 is equal to 10 and 2 more. Repeat with other tens and ones numbers so that pupils become more confident with recognising the 1 ten and some ones without counting ten objects.

- When pupils are confident, the order of the ones and tens can be reversed to demonstrate that 2 ones and 1 ten is also equal to 12. Once pupils are confident with this, they should complete Question 2 in the Practice Book.

10	2
12	

All say... 1 ten and 4 more is equal to 14.

Look out for … pupils who are not yet fluent with teens numbers, for example write 41 instead of 14.

Same-day intervention

- Develop familiarity of matching numerals to quantities through counting out objects into 10 and some more and then matching appropriate arrow cards to the quantities. Combine and say: *10 and 4 more is called 14.*

Same-day enrichment

- Set missing number questions using the bar models.

16
10 + ?

?
10 + 4

17
? + 10

Question 3

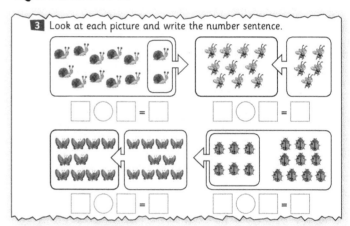

What learning will pupils have achieved at the conclusion of Question 3?

- Pupils will have developed fluency in writing numerals to represent tens and ones numbers.
- Pupils will have begun to use the word 'partition' when explaining how numbers can be separated into tens and ones.
- Pupils will have learned to count on rather than count all to find a total.

Activities for whole-class instruction

- Practise counting on as a whole class. Set out five objects, count them and write the numeral. Set out additional objects and work out the total, counting on from 5. Repeat with other quantities of objects. Use a number line to record the calculation.

- Using a visualiser or concrete objects, lay out ten objects. Check that pupils can identify that there are ten objects. Add two more objects and ask: *How many are there*

Chapter 3 Numbers up to 20 and their addition and subtraction

Unit 3.2 Practice Book 1A, pages 80–81

now in total? It should not be necessary to count all of the objects, but rather count on from 10. Repeat with other quantities between 11 and 20. Increased fluency is achieved when pupils do not need to count on, but can use their knowledge of the structures of the numbers to answer the questions.

- Using a visualiser or concrete objects, lay out 12 objects. Establish that 12 is 10 and 2 more and physically partition into two sets. Remove 2 from this group to illustrate how 12 can be partitioned and ask how many are left. Add 2 to 10 and ask now how many there are and how did you know (without counting)? Repeat with different quantities between 11 and 20.

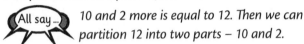

- Pupils should complete Question 3 in the Practice Book.

 10 and 2 more is equal to 12. Then we can partition 12 into two parts – 10 and 2.

Look out for … pupils counting all and not using their knowledge of the structure of the numbers to aid their calculation.

Same-day intervention

- Practise counting on from a number. Count out 10 objects. Rearrange them (into rows, small groups, spread out, bunched up) and check that each time there are still ten objects. Practise adding more objects on to this group of 10, starting the count at 11.
- Count out 10 and some more objects.
- Group the objects into 10 and some more and then physically subtract the ones.

Question 4

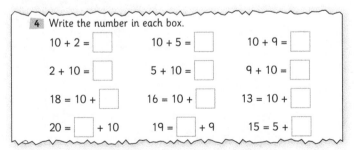

4 Write the number in each box.

10 + 2 =   10 + 5 =   10 + 9 =
2 + 10 =   5 + 10 =   9 + 10 =
18 = 10 +   16 = 10 +   13 = 10 +
20 =   + 10 19 =   + 9   15 = 5 +

What learning will pupils have achieved at the conclusion of Question 4?

- Pupils will have developed greater fluency in matching a numeral to a quantity.
- Pupils will have begun to use the word 'partition' when explaining how numbers can be separated into tens and ones.
- Pupils will have developed their fluency in partitioning numbers into 10 and some more and will apply this to addition and subtraction questions.

Activities for whole-class instruction

- This time, the focus is on written recording and fluency (instantly recalling the answer because they are fluent with partitioning, rather than calculating each answer).

1. Each pupil has a set of arrow cards laid out in front of them. Ask pupils a series of questions and for each one, pupils need to hold up the appropriate arrow card(s), for example: *Show me the number that is equal to 10 and 2 more; Show me the number that is equal to 5 and 10 more; Show me the number that is equal to 3 less than 13; Show me the number that is equal to 14; Show me the number that is equal to …*

2. Write out the number sentences to match each question and answer varying the order of the numbers (total = addend + addend, as well as addend + addend = total).

(i) Arrow cards are an excellent resource for supporting conceptual understanding of place value. '10 and some more' is represented very clearly by arrow cards. By giving pupils opportunities to use arrow cards to physically construct numbers for themselves, you will help them develop and strengthen robust number concepts.

300
60   364
4

- Pupils should complete Question 4 in the Practice Book.

Look out for … pupils still using their fingers or other apparatus to help them count on the numbers that are 10 and some more (rather than being able to use the arrow cards or simply recall these number facts).

Chapter 3 Numbers up to 20 and their addition and subtraction   Unit 3.2   Practice Book 1A, pages 80–81

Same-day intervention

- Practise further with the arrow cards matching task, including starting with cards that represent the quantity visually. Once pupils are confident with matching quantities to numerals, move on to the partitioning questions that focus on the structure of tens and ones numbers to ensure that pupils also become fluent with the language of '10 and some more is equal to xxxxteen'; 'can be partitioned into xxxx and 10'.

Same-day enrichment

- Look at numbers up to 20 in a range of contexts. Using a selection of items for sale and prices between 11p and 19p, pupils should be able to identify (a) that the price 15p is 10p and 5p more and (b) select the appropriate coins to buy that item.

Challenge and extension question

Question 5

5 Write the missing numbers in the boxes.

10 + ☐ = 14   10 + ☐ = 17   ☐ + 1 = 11

☐ + 8 = 18   6 + ☐ = 16   ☐ + 10 = 20

12 − ☐ = 10   16 − ☐ = 10

This question focuses on fluency and being able to use knowledge and understanding of the structure of tens and ones numbers to solve a range of questions. The questions involve combining parts to make the whole, as well as using pupils' understanding of the structure of the numbers to find the value of a missing part when the other part and total are known. This is a precursor to pupils' work on subtraction.

Chapter 3 Numbers up to 20 and their addition and subtraction

Unit 3.3
Ordering numbers up to 20

Conceptual context

This lesson builds on the previous two in which pupils increased their familiarity with quantities of objects and counting. It focuses on developing fluency with using the symbolic notation for numbers as a basis for comparing and ordering numbers. In this unit, the number line is a key model used to develop this knowledge and apply it to different representations. Pupils' familiarity with numbers is further developed and extended through the inclusion of number patterns, comparing the position of numbers and jumbling the order of numbers. This is to ensure that pupils' knowledge of the number system is not limited to rote recitation of number names in order.

Learning pupils will have achieved at the end of the unit

- Pupils will have developed greater fluency in writing and recognising numerals between 1 and 20 (Q1, Q2, Q3, Q4, Q5)
- Pupils will have developed greater fluency in recognising the relative position of numbers on a number line (and how close a number is to a tens number) (Q1, Q3)
- Pupils will have developed their use of a number line to help them visualise the relative size and position of numbers (Q1, Q3)
- Pupils will have developed their ability to compare and order numbers and to spot and describe patterns and relationships with greater fluency (Q1, Q2, Q3, Q4, Q5)
- Pupils will have developed fluency with counting in different-sized steps (twos and fives) (Q2, Q4)

Resources

pictures of quantities from 1 to 20 to accompany numeral cards on a number line (or washing line); arrow cards; mini whiteboards; counters; 1–20 number cards; images of numbers in the real world (such as house numbers)

Vocabulary

ten, 10 and some more, tens and ones, number names (eleven, twelve, thirteen, fourteen, fifteen, sixteen, seventeen, eighteen, nineteen, twenty), odd number, even number, greater than, less than, is equal to, before, after

Chapter 3 Numbers up to 20 and their addition and subtraction Unit 3.3 Practice Book 1A, pages 82–84

Question 1

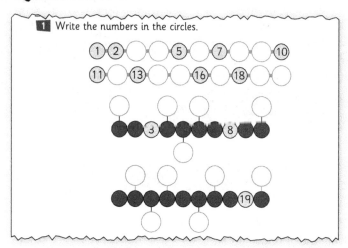

What learning will pupils have achieved at the conclusion of Question 1?

- Pupils will have developed greater fluency in writing and recognising numbers between 1 and 20.
- Pupils will have developed greater fluency in recognising the relative position of numbers on a number line (how close a number is to a tens number).
- Pupils will have developed their ability to compare and order numbers and to spot and describe patterns and relationships with greater fluency.

Activities for whole-class instruction

- Pupils play 'Washing line number line jumble'. In order to develop pupils' familiarity with comparing and ordering numbers, prepare a washing line number line so that some numbers are missing and some are in the wrong order.
- Give some pupils some additional cards that have the number written in words (eleven, twelve, thirteen … up to twenty) and also show the number when it is partitioned, in numerals, for example 10 + 1, 10 + 2.
- As a whole class, begin to count along the number line until you notice there is a problem. Ask pupils to explain what the problem is and to find the correct numeral card. Expect them to use appropriate mathematical vocabulary to provide reasons for the inclusion of a number, such as: *I know that 4 is 1 more than 3 and 2 more than 2 so it must go here next to the 3*. Repeat until the number line from 1 to 20 is complete. Now ask pupils to match the extra cards they have to the numbers on the number line to consolidate the quantity that each number represents.

- Once the number line is complete, ask questions such as: *Tell me what number is 1 less than 15; Tell me what number is 1 more than 17; Tell me what number is 3 more than 12* and so on. Ask pupils to complete Question 1 in the Practice Book.

Same-day intervention

- Use 1–20 number cards. Pupils turn one card over, read the number and if the number is between 11 and 20, partition it. They turn over the next card, read the number and say whether it comes before or after the previous number and whether it is greater than or less than the previous number. If necessary, use pictures of the quantities to help pupils to visualise the quantity represented by the number. Repeat until the whole stack of cards has been turned over and then place the cards in numerical order.

Same-day enrichment

- Look at tens and ones numbers in real-life contexts. Using photos of door numbers, put the numbers in order. Do pupils know their house/flat numbers? Order any that are up to 20. What dates are pupils' birthdays? Can they order those?

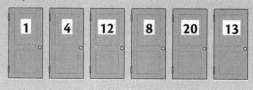

 … pupils who are not yet certain of the name of the next (or previous number).

Question 2

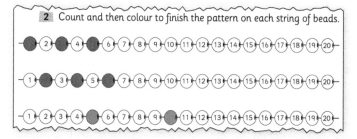

What learning will pupils have achieved at the conclusion of Question 2?

- Pupils will have developed their ability to compare and order numbers and to spot and describe patterns and relationships with greater fluency.

Chapter 3 Numbers up to 20 and their addition and subtraction

Unit 3.3 Practice Book 1A, pages 82–84

- Pupils will have developed fluency with counting in different-sized steps (twos and fives).

Activities for whole-class instruction

- Pupils complete an 'Odd one out' task. The focus of this task is to develop pupils' perceptual abilities and mathematical reasoning through focusing on the characteristics of numbers and the relationships between them.
- Put up groups of three numerals on the board (for example 3, 4, 5). Pupils should work with a partner to explain why each of the numbers could be the odd one out. (For example, 4 is the odd one out because it is the only even number; 5 is the odd one out because it is the only odd number greater than 4; 3 is the odd one out because it is the only number less than 4.) Repeat with different sets of numbers less than 20. Emphasise the importance of the correct use of mathematical terms and develop the set of 'All say' sentences as appropriate.
- Pupils complete Question 2 in the Practice Book.

Look out for … pupils who are not yet fluent with mathematical terms or who cannot identify numbers that are similar or are odd ones out.

Same-day intervention

- Use resources such as same-coloured counters or cubes to make quantities. Arrange them so that it is easier for pupils to see if one quantity is greater than or less than another, or whether or not it is an even or odd quantity.

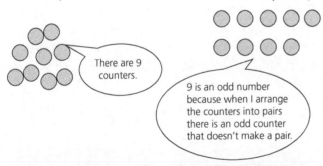

There are 9 counters.

9 is an odd number because when I arrange the counters into pairs there is an odd counter that doesn't make a pair.

Same-day enrichment

- In pairs, one pupil uses some number cards to make a pattern (such as odd numbers, even numbers, every third counting number, every fifth counting number).
- Their partner selects a number card that they think would form part of that pattern and explains why. Once the pattern has been successfully described, the second pupil makes a new number pattern for their partner to continue and describe.

Question 3

3 Write the numbers in the boxes.

☐ 11 ☐ ☐ 14 ☐ ☐ ☐ 18 ☐ ☐

(a) The two numbers before and after 15 are ☐ and ☐.
(b) The number after 18 is ☐ and the number before 11 is ☐.
(c) The number between 11 and 13 is ☐.
(d) The numbers greater than 12 but less than 18 are ☐.
(e) The numbers less than 20 but greater than 15 are ☐.

What learning will pupils have achieved at the conclusion of Question 3?

- Pupils will have developed greater fluency in writing and recognising numerals between 1 and 20.
- Pupils will have become more fluent at identifying and describing patterns and relationships.

Activities for whole-class instruction

- Take a set of number cards from 1 to 20 and turn them face down. Turn over one card and say its value and then partition it to identify that it is 1 ten and some ones.
- Turn over a second card and ask if this number comes before or after the first, and if it is greater than or smaller than. Ask pupils to explain how they know (for example the first number is 10 and 4 more, the second is 10 and 6 more, there are more ones in 16 so 16 is greater than 14).
- Ask pupils to say the names of other numbers that would be between the two digits. Check their answers against a number line.
- Repeat to develop fluency in identifying the larger and smaller numbers.
- Pupils complete Question 3 in the Practice Book.

Same-day intervention

- Pupils turn over a number card and match its position to the same number on a number line. They partition the number and, if necessary, match an image of the quantity to the numeral too.
- They turn over a second card and use the number line to identify whether the second number is before or after the first. Use the images of each quantity to help consolidate which number is greater than or less than the other.

Chapter 3 Numbers up to 20 and their addition and subtraction   Unit 3.3 Practice Book 1A, pages 82–84

Same-day enrichment

- The task in Question 3 required pupils to identify numbers from clues provided in words (the number after 18 is ...). In order to develop fluency in describing the relative position of numbers, pupils work together in pairs to explain how a series of number cards from 1 to 20 can be ordered.
- Start by laying out a set of eight number cards between 1 and 20 in a random order. Position the cards in the correct order then, in order from low to high numbers, ask for a number card to help complete the number line from 1 to 20. For example, using the set of numbers below a pupil may say: *Give me the number that is greater than 10 but less than 12*, or *Give me a number greater than 16*.

| 1 | 5 | 6 | 10 | 12 | 15 | 16 | 20 |

Look out for ... pupils still using their fingers or other apparatus to help them count on the numbers that are 10 and some more (rather than being able to recall these number facts).

Same-day intervention

- In order to help pupils to develop fluency with instant recall of number facts, ask them to play a matching game, using cards that show both the numeral and the quantity. This will help pupils to be able to visualise the 10 and the 'some more'. They should lay out the cards face down and find pairs with equivalent numeral and quantity. As each numeral is matched to the appropriate quantity, ensure that they also become fluent with appropriate mathematical language through describing each quantity as, for example *10 and 5 more is equal to 15 and if we remove 5 from 15 we have 10*.

Question 4

> 4 Swap the positions of two numbers in each set so the new order of the numbers forms a pattern. One has been done for you.
> (a) 1, 3, 9, 7, 5   New: 1, 3, 5, 7, 9
> (b) 20, 19, 18, 17, 15, 16   New: _____
> (c) 3, 6, 5, 4, 7   New: _____
> (d) 4, 12, 8, 16, 20   New: _____

What learning will pupils have achieved at the conclusion of Question 4?

- Pupils will have become more fluent at identifying and describing patterns and relationships.
- Pupils will have developed fluency with counting in different-sized steps (twos and fives).

Activities for whole-class instruction

- Pupils play 'Odd or even?' Start with a set of 0–20 number cards face down. Turn over one card and ask pupils if it is odd or even. If it is even, pupils should say it is even because it is, for example, six groups of 2. If it is odd, pupils should say that it is 1 more than (or 1 less than) an appropriate even number.
- Once five cards have been turned over, put them into numerical order (smallest to largest and then largest to smallest). Reinforce the language of 'is greater than' and 'is less than' by comparing the five cards that have been selected.

Same-day enrichment

- Pupils play 'Odd one out, new one in'. Working in pairs, pupils shuffle a set of 1–20 number cards. They turn over the top three to reveal the numbers. From the set of three numbers, pupils decide which of the numbers could be the odd one out and explain to their partner why, for example: *Two of the numbers are odd and one is even, the even number is the odd one out*, or *They are all odd, but these two are less than 7*.
- Pupils remove the number that is the odd one out and select a new number that they can add in. Pupils should be able to describe the numbers in the sequence they have made, for example: *These are all odd numbers, these all have the 5 digit in them*.

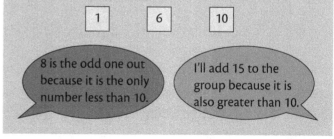

Chapter 3 Numbers up to 20 and their addition and subtraction   Unit 3.3 Practice Book 1A, pages 82–84

Question 5

> 5 Find out and then circle the number in each set below so the remaining numbers form a pattern. One has been done for you.
>
> (a) 3, 5, 7, ⑧, 9   Remaining: 3, 5, 7, 9
> (b) 2, 4, 6, 8, 9   Remaining: _____
> (c) 18, 17, 15, 13, 11   Remaining: _____
> (d) 6, 9, 12, 13, 15, 18   Remaining: _____

What learning will pupils have achieved at the conclusion of Question 5?

- Pupils will have developed greater fluency in writing and recognising numerals between 1 and 20.
- Pupils will have developed their ability to compare and order numbers and to spot and describe patterns and relationships with greater fluency.

Activities for whole-class instruction

- Pupils play 'What's the pattern?' Each pair of pupils has a set of 1–20 number cards. Start by displaying to pupils the first, second and third number in a sequence. In pairs, pupils discuss what they think the pattern is and select numbers from number digit cards that they think would be included in the pattern.
- Then select two numbers from those suggested by pupils that are correct and add at least one further number that may include a number that is not part of the pattern. Add the new numbers to the pattern and ask pupils, in pairs, to describe the pattern and decide if all of the new numbers are part of the pattern. If they are not, they should explain what is different about the numbers that should not have been included.

Same-day intervention

- Being able to describe the mathematical characteristic of numbers in a pattern requires pupils to be able to identify the pattern and use mathematical terms fluently enough to describe it. In order to make this task more accessible, pupils will initially be presented with some patterns to continue that are made up of shapes and colours. When describing the patterns, pupils can use the shape or colour name as well as rehearsing phrases such as 'every other one is', or 'all of these are' 'after' square is circle rather than saying 'it goes square, circle, square, circle'.

- Say to pupils: *Continue and describe the pattern. Say what you think the tenth one will be in each pattern.*

- Once fluent with shape patterns, re-introduce pupils to very simple number patterns.

Same-day enrichment

- Pupils play 'Concept cartoon'. Pupils are presented with some statements about a collection of numbers. They must discuss and agree which statements in the speech bubbles are true and which are false and why.

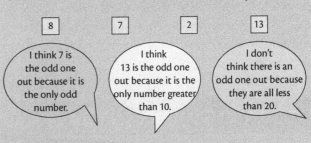

Challenge and extension question

Question 6

> 6 (a) A number is greater than 12 but less than 15.
>   The number could be ☐.
>
> (b) A number is less than 20 but greater than 16.
>   The number could be ☐.
>
> (c) Add some numbers before and after 9 to make a number pattern.
>
>

This question enables pupils to apply what they have learned during this unit. They will become more fluent in the language of comparing and ordering numbers as well as in describing and developing patterns. If you need to extend the question further, ask pupils to write similar questions of their own.

Chapter 3 Numbers up to 20 and their addition and subtraction

Unit 3.4
Addition and subtraction (1)

Conceptual context

The previous three lessons have focused on developing pupils' conceptual understanding of the structure of numbers between 10 and 20 and how applying this knowledge can help to increase fluency with addition and subtraction rather than relying on counting. In order to achieve this, pupils have had many opportunities to increase their familiarity with these numbers as the numbers have been presented using a wide range of physical and pictorial resources, linked consistently to numerals. Although visual representations continue to be part of the teaching for the whole-class activities, the emphasis moves to symbolic notation and encouraging pupils to use instant recall of known facts and identifying and using patterns within calculations to aid in their solution.

Learning pupils will have achieved at the end of the unit

- Pupils will have developed greater fluency in solving simple addition and subtraction questions by applying their knowledge of partitioning numbers (Q1, Q2, Q3, Q4)
- Pupils will have developed flexibility in solving addition and subtraction problems presented through experience with various representations (Q1, Q2)
- Pupils will have reinforced their awareness of number patterns and will have used this to solve problems with greater fluency (Q1, Q2, Q3, Q4)
- Pupils will have investigated patterns within simple pairs of addition and subtraction calculations and applied mathematical reasoning about those patterns to justify their answers to problems (Q2, Q3, Q4)

Resources

ten-frame; coloured flat number tiles; counters; arrow cards; 1–20 number cards; operation cards; chairs; 0–20 number line; frog and lily pads (to stick onto and move along the number line); mini whiteboards

Vocabulary

partition, subtrahend, minuend, addend, is equal to, '10 and some more'

Chapter 3 Numbers up to 20 and their addition and subtraction Unit 3.4 Practice Book 1A, pages 85–87

Question 1

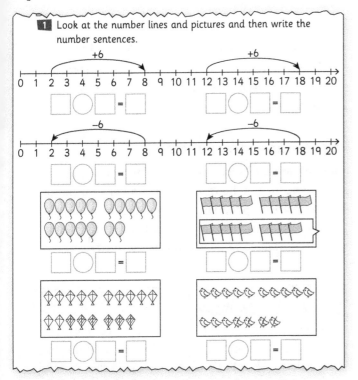

What learning will pupils have achieved at the conclusion of Question 1?

- Pupils will have developed greater fluency in solving simple addition and subtraction questions by applying their knowledge of partitioning numbers.
- Pupils will have developed greater flexibility in solving addition and subtraction calculations between 10 and 20 by solving problems presented using a variety of representations.
- Pupils will have reinforced their awareness of number patterns and will have used this to solve problems with greater fluency.

Activities for whole-class instruction

- Pupils practise jumping forwards and backwards in twos. Use a 0–20 number line with a 'frog' who jumps along it. Starting from 0, move the frog to number 2 and stick on a lily pad to show where he landed. Count on two more jumps of 2. Ask pupils what size jumps the frog has taken and explain that he is counting in twos. Say the first three numbers (2, 4, 6) and then all together count in twos leaving a lily pad on each number as you count. Then count backwards in steps of 2 to help develop fluency.

- Ask pupils if the numbers are odd or even and confirm that they are even. Ask pupils where you would start counting to count up in twos and land only on odd numbers. Starting at 1 show the frog's jumps in twos – hopping from 1 to 3 to 5. Say together the first three numbers (1, 3, 5) and then all together count in twos, leaving a lily pad on each number as you count. Then count backwards in steps of 2 to help develop fluency.

- Pupils practise jumping forwards and backwards in threes. Repeat the frog task, this time demonstrating that the frog lands on every third number. Rehearse saying the counting forwards and backwards, pointing out the numbers for a pattern of odd – even – odd – even.

- Pupils practise jumping forwards and backwards in fives. Repeat the frog task, this time demonstrating that the frog lands on every fifth number. Rehearse saying the counting forwards and backwards, pointing out the numbers for a pattern of odd – even – odd – even.

- How many dots (1)? This task is to help pupils to develop a sense of 'fiveness' both in terms of a quantity and how it can be presented in a variety of different arrangements. The arrangements will help pupils to maintain and develop their understanding of the conservation of number (that though the five dots may be arranged differently, the quantity of dots doesn't change) and also draw on their experience of subitising. Turn over each card in turn; ask pupils how many dots there are and whether they counted, subitised or did a combination of the two.

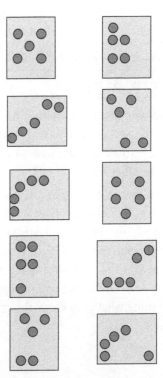

Chapter 3 Numbers up to 20 and their addition and subtraction

Unit 3.4 Practice Book 1A, pages 85–87

Turn over the first of the second row of dotty cards. Ask how many dots there are, this time focusing on whether the quantity is 1 more or 1 less than 5. Turn over other cards.

- How many dots (2)? To help develop the pupil's ability to subitise further, next present fives and some quantities that are 1 more or less than 5 using ten-frames. The arrangement of the counters will help to develop pupils' familiarity with 5 as well as it being equal to 3 + 2, 4 + 1, 3 + 1 + 1 and so on.

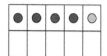

- Using manipulatives, partition quantities between 12 and 19. In order to model some addition and subtractions sentences: first partition them into tens and some more, then partition the ones into two parts where there are more than 5 ones, corresponding with the ten-frame structure. Reinforce symbolic notation through using arrow cards to match the quantity shown. Ensure that pupils are subitising rather than counting all of the counters. Write 10 + 6 = 16; say: *16 is 6 more than 10*; write 16 = 10 + 6; say *16 is 10 more than 6*.

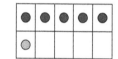

- Point out that in the ten-frame, the 6 is partitioned into 5 and 1 more, and write that as a number sentence (16 = 10 + 5 + 1 = 10 + 1 + 5).
- Point out to pupils that this could also be written as 16 = 15 + 1 = 11 + 5. Show this on a number line.

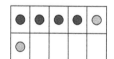

- Rearrange the 6 ones and reinforce that there are still 6 ones and write the new corresponding number sentences (16 = 10 + 4 + 2 = 10 + 2 + 4 and 16 + 14 + 2 = 12 + 4). Show this on a number line.
- Once pupils have become fluent in recognising the addition sentences focus on the inverse calculations, for example 16 – 2 = ? 16 – 14 = ? Model this in a variety of ways such as removing the quantity to be subtracted and crossing out the quantity to be subtracted, or circling the subtrahend. Ask pupils the answer, emphasising the importance of using what they know about the numbers in the calculation.

- Use a number line to show the inverse relationships between the numbers in the calculations: 14 + 2 = 16, so 16 – 2 = 14.
- Pupils complete Question 1 in the Practice Book.

Look out for ... pupils counting in ones and not recognising the relationships between the numbers in the calculations.

Same-day intervention

- Use a resource such as coloured flat number tiles to explore further the structure of numbers. Lay out one 10 tile, then a 6 tile. Use the other tiles to show pairs of numbers equal to 6. For each pair, say: *xxxteen added to yyyy is equal to 16; 16 take away yyyy is equal to 10.*

Same-day enrichment

- To develop their familiarity and fluency with using known facts to solve simple addition and subtraction questions, ask pupils to work systematically to find all of the possible number bonds that could be added to 10 to equal 16. Highlight that some pairs of numbers (5 + 1 and 1 + 5) remind us that addition is commutative and the order in which the numbers are added does not change the total.

Question 2

2 Complete the calculations.

6 + 2 = ☐	3 + 5 = ☐	4 + 4 = ☐	5 + 2 = ☐
16 + 2 = ☐	13 + 5 = ☐	14 + 4 = ☐	15 + 2 = ☐
1 + 5 = ☐	6 + 3 = ☐	3 + 3 = ☐	2 + 4 = ☐
11 + 5 = ☐	16 + 3 = ☐	13 + 3 = ☐	12 + 4 = ☐

What learning will pupils have achieved at the conclusion of Question 2?

- Pupils will have developed greater fluency in solving simple addition and subtraction questions by applying their knowledge of partitioning numbers.

Chapter 3 Numbers up to 20 and their addition and subtraction

Unit 3.4 Practice Book 1A, pages 85–87

- Pupils will have investigated patterns within simple pairs of addition and subtraction calculations and applied mathematical reasoning about those patterns to justify their answers to problems.
- Pupils will have developed greater flexibility in solving addition and subtraction calculations between 10 and 20 by solving problems presented using a variety of representations.

Activities for whole-class instruction

- Use coloured flat number tiles to model adding together two single-digit numbers, writing the addition sentence in numerals below each one and using arrow cards for each numeral. Pupils should be able to answer rapidly questions such as 5 + 3, 4 + 5, 1 + 7, 6 + 2, 3 + 4 without counting.
- For each calculation, change the first addend to 10 more (15 + 3, 14 + 1 …). Model the calculation by including the 10 coloured flat number tile and the 10 arrow card. Ask pupils: *What is the same? What is different?* and encourage them to use what they know to answer each calculation.

 If 5 + 3 is equal to 8, then 15 + 3 is equal to 18 because 15 is 10 more than 5.

 … pupils who are counting on rather than understanding the effect of adding 10 to the addend.

- Ask pupils to complete Question 2 in the Practice Book, drawing their attention to the pattern in the questions.

Same-day intervention

- Provide additional opportunities to develop pupils' ability to subitise using ten-frames. Always 'fill' the frame one row of 5 at a time, from left to right, before starting to fill the next row.

- Double-sided counters are particularly helpful. Ask pupils to represent 6 + 3 = .

Now ask them to add 10.

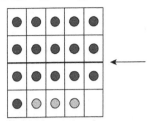

- Can pupils see that by adding 10 to the addend, the sum increases by 10?
- Repeat with: *Start by adding two small single-digit numbers to make a single-digit sum – add 10.*
- Pupils complete Question 2 in the Practice Book.

Same-day enrichment

- Finding all of the possibilities and working systematically, pupils will find all of the pairs of numbers that total 8 and then all of the pairs of numbers that total 18. Pupils will be able to justify that they have included all of the possibilities and be able to explain that there are pairs of calculations that give the same answer regardless of the order the numbers are added, and use resources such as cubes or ten-frames to demonstrate that: 1 + 7 = 7 + 1, 8 = 1 + 7 = 2 + 6 = 3 + 5 = 4 + 4 = 5 + 3 = 6 + 2 = 7 + 1

Question 3

3 Complete the calculations.

6 – 2 = ☐   7 – 5 = ☐   4 – 3 = ☐   5 – 2 = ☐

16 – 2 = ☐   17 – 5 = ☐   14 – 3 = ☐   15 – 2 = ☐

9 – 5 = ☐   6 – 4 = ☐   8 – 3 = ☐   9 – 4 = ☐

19 – 5 = ☐   16 – 4 = ☐   18 – 3 = ☐   19 – 4 = ☐

What learning will pupils have achieved at the conclusion of Question 3?

- Pupils will have developed greater fluency in solving simple addition and subtraction questions by applying their knowledge of partitioning numbers.
- Pupils will have reinforced their awareness of number patterns and used this to solve problems with greater fluency.

173

Chapter 3 Numbers up to 20 and their addition and subtraction   Unit 3.4 Practice Book 1A, pages 85–87

- Pupils will have investigated patterns within simple pairs of addition and subtraction calculations and used mathematical reasoning to justify their answers.

Activities for whole-class instruction

- In order to make explicit connections between addition and subtraction facts and reinforce the relationships between part–whole in addition and subtraction, repeat the previous task, this time focusing on subtraction sentences.
- Ensuring that the minuend is always greater than the subtrahend, use bars to model subtracting two single-digit numbers, writing the subtraction sentence in numerals below each one, for example 10 – 6 = 4, then 20 – 6 = 14.

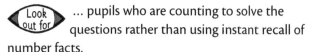

- For each calculation, change the minuend to 10 more (20 – 7, 18 – 3 …). Model the calculation through including the double ten-frame and bar models. Ask pupils: *What is the same/What is different about the questions?* Make explicit that the second set of numbers comprise 10 and some more and encourage pupils to use what they know to answer each calculation.
- Pupils should complete Question 3, using known facts to answer questions rapidly when appropriate.

Look out for … pupils who are counting to solve the questions rather than using instant recall of number facts.

Same-day intervention

- Pupils should continue to use ten-frames to create tens and ones numbers in order to develop instant recall of simple addition and subtraction facts. Start with a teens number and subtract 10, reinforcing the language that 14 is equal to 10 and 4 more so 14 subtract 10 is equal to 4. Ensure that pupils also use the ten-frames to subtract 10 from a teens number, and recognise that the difference between the minuend and the subtrahend is the number of ones.

Same-day enrichment

- Finding all of the possibilities and working systematically, pupils should find all of the pairs of numbers less than 20 with a difference of 3. Pupils will be able to justify that they have included all of the possibilities and be able to explain that the order of the numbers is important – that, for example 8 – 5 is not equal to 5 – 8.

Question 4

4 Think carefully and then fill in the boxes.

3 + 5 = ☐   5 + 3 = ☐   2 + 7 = ☐

13 + 5 = ☐   ☐ + ☐ = ☐   ☐ + ☐ = ☐

4 – 2 = ☐   6 – 5 = ☐   7 – 2 = ☐

14 – 2 = ☐   ☐ – ☐ = ☐   ☐ – ☐ = ☐

What learning will pupils have achieved at the conclusion of Question 4?

- Pupils will have developed greater fluency in solving simple addition and subtraction questions by applying their knowledge of partitioning numbers.
- Pupils will have reinforced their awareness of number patterns and will have used this to solve problems with greater fluency.
- Pupils will have investigated patterns within simple pairs of addition and subtraction calculations and used mathematical reasoning to justify their answers.

Activities for whole-class instruction

- Pupils should be becoming increasingly fluent with their addition and subtraction calculations. They should be drawing on known facts to provide immediate recall for questions and be able to derive their own questions using facts they can instantly recall.
- Play 'How many questions can you make?' Write a number between 3 and 20 on the board. In mixed attainment pairs, pupils must provide as many addition and subtraction questions as they can within one minute and record these on their mini whiteboards.

Chapter 3 Numbers up to 20 and their addition and subtraction

Unit 3.4 Practice Book 1A, pages 85–87

Look out for ... pupils who are treating 7 − 3 as equal to 3 − 7 because mentally they are taking the smaller number away from the larger number rather than subtracting the subtrahend from the minuend.

Same-day intervention

- Pupils turn over two digit cards less than 20. They identify which one is greater and which one is smaller. They then arrange the numbers to make a subtraction sentence and solve the calculation using instant recall (use a number line or objects, if necessary). Pupils rehearse reading the subtraction sentence in the order that it is written and try to use instant recall of subtraction and related addition facts to find and check the answer.

Same-day enrichment

- Provide pupils with an opportunity to develop their mathematical reasoning. Pupils are given some addition and calculation questions with some of the digits covered over. They must say the value of the missing digit and explain how they know what it must be, based on known facts (rather than an explanation of a counting strategy).

  1◊ + 5 = 17   14 + ◊ = 19
  1◊ + 1 = 13   12 + ◊ = 19
  19 − ◊ = 12   1◊ − 5 = 12
  16 − ◊ = 10   19 − ◊ = 15

Challenge and extension question

Question 5

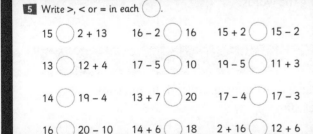

5 Write >, < or = in each ◯.

15 ◯ 2 + 13   16 − 2 ◯ 16   15 + 2 ◯ 15 − 2
13 ◯ 12 + 4   17 − 5 ◯ 10   19 − 5 ◯ 11 + 3
14 ◯ 19 − 4   13 + 7 ◯ 20   17 − 4 ◯ 17 − 3
16 ◯ 20 − 10   14 + 6 ◯ 18   2 + 16 ◯ 12 + 6

The challenge and extension question for this unit develops pupils' understanding of equivalence and their use of the equals and inequalities symbols.

Chapter 3 Numbers up to 20 and their addition and subtraction

Unit 3.5
Addition and subtraction (II) (1)

Conceptual context

This unit focuses on developing pupils' understanding of bridging as a calculation strategy. It revisits numbers that will be familiar to pupils and begins by focusing on calculations that pupils will be able to solve using instant recall. This allows the focus to be on the flexible partitioning that is required when the bridging strategy is used, rather than focusing solely on finding the answer to the calculations. To ensure that all pupils develop both a conceptual and procedural understanding of bridging, teaching continues to be supported through representing the mathematics in a variety of physical, pictorial and symbolic ways. The interaction with images and resources also helps pupils to fully understand the meaning of mathematical terms such as bridging and addend, and recognise how symbolic notation reflects the steps in the calculation.

Learning pupils will have achieved at the end of the unit

- Pupils will have developed their range of addition calculation strategies to include bridging (Q1, Q2, Q3, Q4)
- Pupils will have developed their fluency with partitioning single-digit numbers in order to use the bridging strategy (Q1, Q2, Q3, Q4)
- Pupils will have had further opportunities to partition a small number effectively (Q1, Q2, Q3, Q4)
- Pupils will have reinforced their understanding of bridging through using a variety of resources and representations (Q1, Q2, Q3, Q4)
- Pupils will have become more fluent with using symbolic notation in a variety of ways (Q2, Q3, Q4)
- Pupils will have developed their skills in communicating orally using the words addend and partition (Q1, Q2, Q3, Q4)

Resources
ten-frames; number lines; counters; coloured flat number tiles; digit cards; mini whiteboards

Vocabulary
bridging, partitioning, addend, 10 and some more, total, sum

Chapter 3 Numbers up to 20 and their addition and subtraction

Unit 3.5 Practice Book 1A, pages 88–90

Question 1

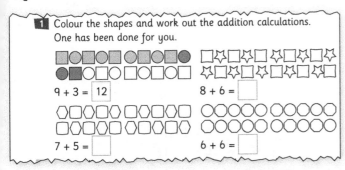

1 Colour the shapes and work out the addition calculations. One has been done for you.

9 + 3 = 12   8 + 6 = ☐

7 + 5 = ☐   6 + 6 = ☐

What learning will pupils have achieved at the conclusion of Question 1?

- Pupils will have developed their range of calculation strategies to include bridging.
- Pupils will have had further opportunities to partition a small number effectively.
- Pupils will have reinforced their understanding of bridging through using a variety of resources and representations.
- Pupils will have developed their fluency with partitioning single-digit numbers in order to use the bridging strategy.
- Pupils will have become more fluent with using symbolic notation.

Activities for whole-class instruction

- These tasks use numbers with which pupils are familiar and model for them how numbers are partitioned when using the bridging procedure for adding. Familiar numbers enable pupils to focus more on the procedure of adding the numbers rather than focusing on working with unfamiliar numbers.
- For partitioning numbers less than 10, for example a number between 3 and 9, and in pairs, pupils write on their whiteboards pairs of numbers with that total.
- Pupils make tens using a ten-frame and a number line. Using two ten-frames, model adding two single-digit numbers making totals greater than 10, for example 7 + 5. Lay out the first addend (7) and show the second addend (5) ready to go into the ten-frame. Ask pupils how many are needed to fill up the first ten-frame and then put the remaining two counters into the second ten-frame. Write out the calculation in steps (7 + 3 = 10, 10 + 2 = 12) highlighting that the second phase is using their instant recall of making numbers between 10 and 20 (10 and 2 more is equal to 12).

Point to the red counters. Say: *This is the 7, the first addend.* Point to the 3 and 2, say: *This is the 5, the second addend. The 5 needed to be partitioned into 3 (to make the 7 up to 10) and then 2, to start the next 10.*

Ask, while pointing again to the two addends: *7 add 5 equals?*

All say... *Partition the second addend to make 10, then add on. This is called bridging.*

- Pupils then complete Question 1 in the Practice Book.

Look out for ... pupils who count on in ones from the first addend (do not use a bridging strategy).

Same-day intervention

- Use coloured flat number tiles to aid calculation of single-digit numbers to make a group of 10 and some more. This should help to encourage pupils not to count on in ones (as they may do if they are laying out counters to complete a ten-frame). For example, for 7 + 5, select the '7' and the '5' tile. Lay the 7 tile on top of a 10 tile and ask the pupil how many more are needed to make 10. Lay a 3 tile and a 2 tile on top of the 5 tile to show that 3 + 2 is equal to 5, then add the 3 to the 7 to make 10, followed by adding the 2 to equal 12.
- Write the calculation in stages. This will help pupils to connect the action of modelling the calculation with the coloured number tiles to the symbolic recording of that action.

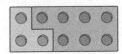

- Use number lines to model counting on in efficient steps. For example, for 7 + 5, use a number line to show the first jump going from 7 to 10 and the second from 10 to 12. Pupils should be discouraged from making jumps of 1, but rather use the facts that they know about number bonds to 10 and partitioning numbers to make more efficient jumps.

Chapter 3 Numbers up to 20 and their addition and subtraction Unit 3.5 Practice Book 1A, pages 88–90

Same-day enrichment

- Provide 'empty box' questions that focus on the partitioning of an addend to bridge through 10. These can be presented as written calculations, on a ten-frame or on a number line. In each case, pupils should identify which number had been partitioned to aid bridging.

  8 + ☐ + 4 = 14 What was the calculation before the addend was partitioned?

  17 = 6 + ☐ + 7 What was the calculation before the addend was partitioned?

  8 + ☐ + 5 = 15 What was the calculation before the addend was partitioned?

Question 2

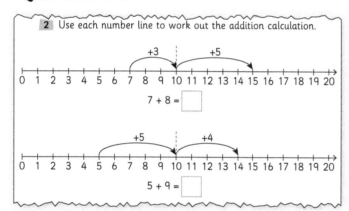

What learning will pupils have achieved at the conclusion of Question 2?

- Pupils will have developed their range of calculation strategies to include bridging.
- Pupils will have had further opportunities to partition a small number effectively.
- Pupils will have reinforced their understanding of bridging through using a variety of resources and representations.
- Pupils will have developed their fluency with partitioning single-digit numbers in order to use the bridging strategy.
- Pupils will have become more fluent with using symbolic notation.

Activities for whole-class instruction

- In order to develop pupils' understanding of the concept of partitioning, making tens and the procedure of bridging, the ten-frame activity described above is revisited. This time, the partitioning and bridging is reinforced through modelling the stages on a number line.

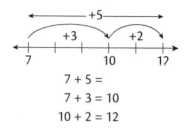

7 + 5 =
7 + 3 = 10
10 + 2 = 12

- Repeat with further pairs of numbers, varying the focus:

  (a) present the ten-frame and ask pupils to draw a number line to match

  (b) from the number line, ask pupils what the calculation is

  (c) from the calculation, ask pupils to say how they partitioned the addend to make 10. You should draw pupils' attention to the adding on from 10 and ensure that they are instantly recalling these facts rather than calculating them.

- Pupils then complete Question 2 in the Practice Book.

  All say... Partition the addend to make 10, then add on. This is called bridging.

  Look out for ... pupils who count on in ones from the first addend (do not use a bridging strategy) or who are counting on using their fingers rather than using instant recall of number facts.

Same-day intervention

- It is important that pupils can move fluently between visual and symbolic representations of calculations so that they are not reliant on counting, and that they fully understand what the symbols within a calculation represent so that they can partition the second addend as efficiently as possible. Pupils should work in pairs to create the steps used when bridging using number lines. At each stage, they should explain to their partner what they are doing.

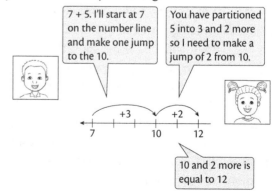

Chapter 3 Numbers up to 20 and their addition and subtraction

Unit 3.5   Practice Book 1A, pages 88–90

Same-day enrichment

- In order to develop greater fluency with using the bridging method, pupils should develop their instant recall of number facts, which they will utilise when adding a single-digit number using the bridging strategy. In pairs, pupils should list all of the different ways in which 4, 5, 6, 7, 8 and 9 can be partitioned into two numbers. (Encourage them to do this systematically and to be aware that addition is commutative; it is important to recognise that 9 is equal to 1 added to 8 as well as 8 added to 1.)
- Pupils begin to generalise about how best to partition the second addend. For example,

7 can be partitioned into:	1 + 6	2 + 5	3 + 4

- For the calculation 8 + 7, it is best to partition the 7 into 2 and 5 more. Can pupils explain why? *Because 2 is what must be added to 8 to get to 10 so I need the number bond for 7 that includes 2.*
- For the calculation 6 + 7, partition the 7 into 4 and 3 more. Can pupils explain why? *Because 4 is what must be added to 6 to get to 10 so I need the number bond for 7 that includes 4.*
- For the calculation 9 + 7, partition the 7 into 9 and 1 more. Can pupils explain why?

- Pupils will have reinforced their understanding of bridging through using a variety of resources and representations.
- Pupils will have developed their fluency with partitioning single-digit numbers in order to use the bridging strategy.
- Pupils will have become more fluent with using symbolic notation.

Activities for whole-class instruction

- This activity will help pupils to record their bridging calculations more efficiently and become more fluent in partitioning small numbers to make 10 when bridging. In pairs, pupils have sets of 1–9 digit cards. On the board, write up the first addend (a number between 3 and 9) and the second addend (a number that would make a total greater than 10). Each pair selects the appropriate digit cards to show how the second addend is partitioned to make 10 and some more.
- Pupils complete Question 3 in the Practice Book.

  All say… Partition the addend to make 10, then add on. This is called bridging.

  Look out for … pupils who are unable to work with symbols fluently and need to count using fingers or along a printed number line to solve the calculations.

Question 3

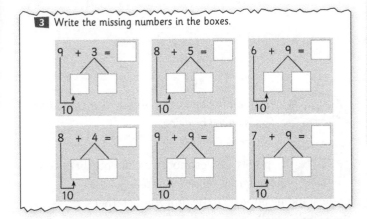

What learning will pupils have achieved at the conclusion of Question 3?

- Pupils will have developed their range of calculation strategies to include bridging.
- Pupils will have had further opportunities to partition a small number effectively.

Same-day intervention

- Repeat the whole-class instruction task, but as well as partitioning the second addend using digit cards, supplement this by representing the question using coloured, flat number tiles. This will help to provide pupils with additional experiences of matching symbols to numerals so that the numerals represent quantities for pupils. The task can also be repeated using number lines and ten-frames.

Same-day enrichment

- In order to become more fluent when recording their bridging calculations, pupils write equivalent number sentences and ask a partner to solve missing number questions presented in number sentences:

  9 + 3 = 9 + 1 + ☐ = 12
  8 + 5 = 8 + ☐ + 3 = 13

Chapter 3 Numbers up to 20 and their addition and subtraction   Unit 3.5   Practice Book 1A, pages 88–90

Question 4

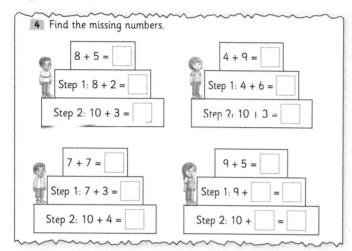

What learning will pupils have achieved at the conclusion of Question 4?

- Pupils will have developed their range of calculation strategies to include bridging.
- Pupils will have had further opportunities to partition a small number effectively.
- Pupils will have reinforced their understanding of bridging through using a variety of resources and representations.
- Pupils will have developed their fluency with partitioning single-digit numbers in order to use the bridging strategy.
- Pupils will have become more fluent with using symbolic notation.

Activities for whole-class instruction

- Pupils have, by now, had much experience of applying the strategy of bridging to subtraction calculations. They have also already used a variety of recording representations: ten-frames, number lines and bridging diagrams. The aim now is to increase fluency with bridging and using written recording as a means of recording the stages rather than as an aid to help identify them.
- This activity will help pupils to record their bridging calculations more efficiently and become more fluent in partitioning small numbers in order to reach a 10 when bridging.
- First, repeat the task from the previous question (working in pairs to identify how the second addend should be partitioned) using a number line to highlight how the second addend is partitioned.
- Next, remind the class of the bridging diagram used in Question 3 and ask pupils to describe the stages. Model rewriting the number line example as a bridging diagram. If pupils understand, give them another calculation to show on a number line and as a bridging diagram. Model further examples as necessary.
- Next, ask them again to describe in words what is happening. Give the starting question 5 + 7 =
  Write:
  5 + 7 = ☐
  Step 1: 5 + ☐ = ☐
- Remind them of what they know about how to bridge – about the 'All say … ' that they have be saying (*Partition the addend to make 10, then add on. This is called bridging*). Record the steps on the calculation together.

  5 + 7 = ☐
  Step 1: 5 + 5 = 10
  Step 2: 10 + ☐ = ☐

- Give the class a question presented using the two-stage model and ask them to draw the corresponding number line. Each time, rehearse orally the steps that they are going through. Ask pupils, *What is the same?* and *What is different?* about each representation. Emphasise that they both show the two steps involved in bridging; partition to make 10, then add on.

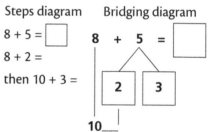

 Partition the addend to make 10, then add on. This is called bridging.

Look out for … pupils who cannot move fluently between the different representations (whose understanding appears to be procedural rather than conceptual).

Chapter 3 Numbers up to 20 and their addition and subtraction   Unit 3.5   Practice Book 1A, pages 88–90

Same-day intervention

- Beginning with the number line representation, ask pupils to explain the steps that are involved in adding two numbers such as 8 + 7. If necessary, help them to support their explanation with the aid of a resource such as a ten-frame or coloured flat number tiles. Ensure that they are using mathematical language correctly. Emphasise the two-step nature of the bridging strategy: make a 10 and then add on the some more.

Same-day enrichment

- In order to develop pupils' fluency with partitioning and with the using the stage-by-stage model, pupils are provided with examples of the stage-by-stage model but the position of the knowns and unknowns will vary.

    8 + 6 =
    8 + ☐ =
    10 + ☐ =

Challenge and extension question

Question 5

5 Use the picture to write addition sentences.

By size: _____

By shape: _____

Up to this point, pupils have been finding the answers to calculations they have been given. Now they have the opportunity to create their own calculations. Using an image showing a variety of objects (such as a bowl of fruit or sweets) ask pupils to suggest their own questions that would involve making totals between 10 and 20. For example, 7 apples and 8 oranges = 7 + 3 + 5 = 15 pieces of fruit.

Chapter 3 Numbers up to 20 and their addition and subtraction

Unit 3.6
Addition and Subtraction (II) (2)

Conceptual context

In the previous unit, pupils learned how bridging can be used as a strategy for adding numbers and developed fluency in partitioning addends to aid their calculations. In this unit, pupils develop their application of calculation strategies that they have learned with numbers up to 20. The unit is underpinned by first-hand experiences to ensure that pupils understand that commutativity and inverse operations are not just about re-ordering numbers but recognising that those numbers represent quantities. Once again, simple single-digit calculations are used so that pupils can focus on the mathematics underpinning the strategy rather than only on the answer itself, since it is the understanding of the process that is most important at this stage. However, it is also important that pupils retain their instant recall of number facts and also maintain a sense of the approximate size of an answer so these are rehearsed to maintain fluency.

Learning pupils will have achieved at the end of the unit

- Knowledge of counting on will have been consolidated (Q1)
- Understanding of commutativity in addition will have been developed (Q1)
- Pupils will have continued to develop a range of calculation strategies and apply these appropriately (Q1, Q2, Q3)
- Pupils will have continued to develop instant recall of addition facts to 20 (Q1, Q2, Q3)
- Use of inequality symbols and ability to approximate the size of an answer will have been consolidated (Q4)
- Pupils will have practised using the bridging strategy (Q3, Q4)
- Pupils will have practised using the words addend, total, commutative and inverse relationships (Q1, Q2, Q3, Q4)
- Pupils will have had further opportunities to explain and justify their decisions using appropriate mathematical language (Q1, Q2, Q3, Q4)

Resources

coloured rods (or home-made coloured cards); counters; coloured flat number tiles; digit cards; number lines; buttons; box with a lid; removable sticky labels; pan balance; cubes; mini whiteboards

Vocabulary

commutative/commutativity, is equal to, addend

Chapter 3 Numbers up to 20 and their addition and subtraction Unit 3.6 Practice Book 1A, pages 91–93

Question 1

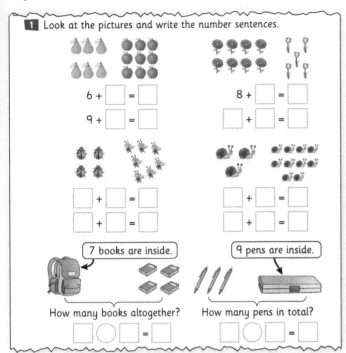

What learning will pupils have achieved at the conclusion of Question 1?

- Knowledge of counting on will have been consolidated.
- Pupils will have continued to develop a range of calculation strategies and apply these appropriately.
- Understanding of commutativity in addition will have been developed.
- Pupils will have continued to develop instant recall of addition facts to 20.
- Pupils will have practised using the words addend, commutative and inverse relationships.
- Pupils will have had further opportunities to explain and justify their decisions using appropriate mathematical language.

Activities for whole-class instruction

- Using rods coloured yellow, green and red, lay out three rods.

    yellow

    green   red

- Ask pupils what they notice about the rods. Remind pupils that in Unit 3.2 they learned that the rods show how the yellow rod is equal to the green rod and the red rod together: the green rod added to the red rod is equal to the yellow rod. Reverse the positions of the green and red rods

- Ask pupils again what they notice about the rods. Emphasise that the yellow rod is equal to the red rod and the green rod together. Explain that this demonstrates that addition is commutative, which means that the order in which you add the numbers does not change the answer.
- Substitute some numbers for the colours. Ask pupils what the answer would be if the green rod was equal to 8 and the red rod was equal to 6. Write out the number sentence 8 + 6 = 14. Ask pupils: *What is the sum of 8 and 6?*
- Clarify that, as addition is commutative, changing the order of the addends does not change the total, and that this knowledge can be useful when trying to add quickly and easily.
- Count out 12 buttons and put them on the desk (or board). Count out a further three buttons. Ask pupils if it would be quicker to add 12 onto 3 or add 3 onto 12.

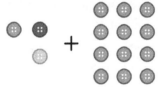

- Write 12 + 3 = 15 = 3 + 12.
- Pupils should then complete Question 1 in the Practice Book.

     Addition is commutative so the order in which you add the numbers does not change the sum.

    ... pupils who are not yet confident with commutativity and think that the order of the numbers in addition changes the sum.

Same-day intervention

- Initially, use counters to create two sets of objects. Pupils count the number in each set, write the addition calculation and work out the sum. Swap over the two sets, write the addition calculation and work out the sum checking that pupils recognise that the sum is not changed by the order in which the numbers are added. Record the calculations on number lines to help further emphasise the equivalence of the two calculations.

Same-day enrichment

- Working in pairs, pupils should plan and rehearse an explanation to a parent/carer about what they have been learning today. They should be ready to 'perform' it with their partner to the class at the end.

Chapter 3 Numbers up to 20 and their addition and subtraction

Unit 3.6 Practice Book 1A, pages 91–93

Question 2

2 Complete the table.

addend	8	9	5
addend	5	4	7
sum			

addend	9	3	6
addend	8	8	5
sum			

What learning will pupils have achieved at the conclusion of Question 2?

- Pupils will have continued to develop a range of calculation strategies and apply these appropriately.
- Pupils will have continued to develop instant recall of addition facts to 20.
- Pupils will have practised using the words addend, commutative and inverse relationships.
- Pupils will have had further opportunities to explain and justify their decisions using appropriate mathematical language.

Activities for whole-class instruction

- Best way round? Once pupils have developed their understanding of commutativity, the focus moves to when it is appropriate to apply that knowledge.
- Write on the board the calculation 4 + 9. Re-introduce the term addend and explain that in this addition calculation there are two addends. An addend is a number that you add to another addend to make a total.
- Ask pupils what the two addends are in the question. Ask them if it is easier to add 9 to 4 or 4 to 9? (Note: some pupils may not discern a difference between the two calculations because they have instant recall of both facts.) Use a number line to demonstrate that adding on 4 to 9 is more efficient than adding 9 to 4 because there are fewer ones to add. Ask pupils to use correct mathematical language to explain why the sums are equal when the order of the addends is changed and reinforce that, as addition is commutative, then changing the order in which the addends are added does not change the sum.
- Vary the representation and present pupils with two ten-frames (representing, for example 13 and 4). Write the two addition sentences that they represent and ask pupils which is the best way round and why.

- Pupils should then complete Question 2 in the Practice Book.

 Addition is commutative so changing the order of the addends does not change the sum.

 … pupils who are not changing the order of the numbers to increase efficiency when calculating.

Same-day intervention

- Using coloured number tiles to create addition sentences, swap around the numbers and ask pupils what they notice about the tiles. Calculate the total and check that the sum is unaltered. Ask pupils whether it is more efficient to count on a small amount or a larger amount, modelling the steps using a number line.

Same-day enrichment

- Give pupils sets of three single-digit numbers, for example 3, 6 and 7. For each set of numbers, they must decide if they should re-order the addends to make the sum easier to solve, or if they should re-order and use a bridging strategy to find the sum. Pupils should be able to justify their decisions drawing on their knowledge of commutativity and number bonds to 10.

Question 3

3 Work out the sums.

9 + 2 =   8 + 3 =   7 + 5 =   4 + 9 =
9 + 3 =   8 + 4 =   7 + 6 =   3 + 8 =
9 + 4 =   8 + 5 =   7 + 7 =   3 + 9 =
9 + 5 =   8 + 6 =   7 + 8 =   2 + 9 =
9 + 6 =   8 + 7 =   7 + 9 =   9 + 7 =
8 + 8 =   6 + 5 =   9 + 8 =   8 + 9 =
6 + 6 =   9 + 9 =   7 + 4 =   6 + 7 =

Chapter 3 Numbers up to 20 and their addition and subtraction   Unit 3.6 Practice Book 1A, pages 91–93

What learning will pupils have achieved at the conclusion of Question 3?

- Pupils will have continued to develop a range of calculation strategies and apply these appropriately.
- Pupils will have continued to develop instant recall of addition facts to 20.
- Pupils will have had further opportunities to practise using the bridging strategy.
- Pupils will have practised using the words addend, commutative and inverse relationships.
- Pupils will have had further opportunities to explain and justify their decisions using appropriate mathematical language.

Activities for whole-class instruction

- It is important that pupils develop and maintain flexible and fluent calculation strategies and are able to apply the most efficient strategy when calculating. To this end, even though this unit focuses on developing pupils' understanding of commutativity and how this can be applied to addition calculations, pupils should also continue to rehearse the bridging strategy learned in the previous unit and begin to be able to select the most appropriate strategies when calculating.
- What size steps? Using a number line, demonstrate adding a single-digit number to a 9. Ask pupils what they notice about the size of the steps each time and clarify that the first step is always a 1 because 9 + 1 = 10. Repeat this with other addends, clarifying that the size of the first step should always make a total of 10. Pupils should be able to generalise that how the second addend should be partitioned depends on how close the first addend is to 10.
- Pupils should then complete Question 3 in the Practice Book.
- Calculation sort. Present the class with a collection of 10 single-digit addition calculations with totals from 11 to 20 presented on separate cards.
- In pairs, pupils must discuss and decide which is the best way to solve the calculation. Do they choose to re-order the numbers, or to leave them as they are and bridge. Do they need to do both? As a class, discuss each calculation, which is the best strategy and why; emphasise the importance of having a range of strategies to help them to calculate efficiently.

 … pupils who always use the same strategy rather than the most efficient strategy.

Same-day intervention

- A pupil's choice of strategy is often based on personal preference, their knowledge of and confidence with number bonds, and the relationship between numbers. In order to develop pupils' confidence with bridging, it is important that they are very familiar with how close a number is to a multiple of 10. In order to help develop this, ten-frames can be used to consolidate pupils' understanding of how close a number is to a 10. Using a set of 1–9 digit cards, select one card.
- Pupils should add red counters to the ten-frame to match the numeral and then yellow counters to model how close the number is to 10 (for example 7 red counters and 3 yellow ones as 7 is 3 less than 10).

- Once pupils are working confidently with numbers up to 10, they can then work with numbers from 11 to 19.
- Pupils should add red counters to the ten-frame to match the numeral and then yellow counters to model how close the number is to 20 (for example 17 red counters and 3 yellow ones as 17 is 3 less than 10).

- Use materials and images such as number lines, counters and coloured flat number tiles to model simple addition calculations. For each one, read the question, establish which addend is greater, and whether it is more efficient to add on a larger or a smaller addend. Identify whether the greater addend is close to 10 and use number lines to clarify how the second addend can be partitioned to make a 10 and some more.

Chapter 3 Numbers up to 20 and their addition and subtraction   Unit 3.6 Practice Book 1A, pages 91–93

Same-day enrichment

- Calculation maker. Give pupils three sets of 1–9 digit cards and a set of strategy cards that say 're-order', 'bridging' or 're-ordering and bridging'. In pairs, one pupil selects a strategy card and their partner must select one digit card from each pile to make a three-digit addition calculation to match the strategy selected. They have to justify their decision. If the pupil has chosen numbers that suit that strategy then they score a point, if their partner can explain a more efficient strategy for solving that calculation then they score a point. The roles are then swapped for the second calculation. The winner is the first to score 10 points.

Question 4

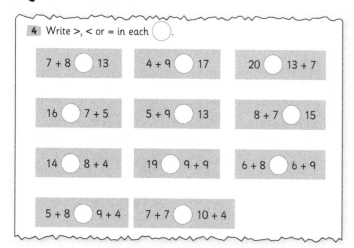

What learning will pupils have achieved at the conclusion of Question 4?

- Pupils will have consolidated their use of 'greater than' and 'less than' symbols and approximating the size of an answer.
- Pupils will have had further opportunities to practise using the bridging strategy.
- Pupils will have practised using the words addend, commutative and inverse relationships.
- Pupils will have had further opportunities to explain and justify their decisions using appropriate mathematical language.

Activities for whole-class instruction

- Get the balance right. Ask five pupils to sit on a row of chairs. Ask the pupil in the fourth position to hold a <, > or = sign. Give pupils on the left of the pupil two numbers and an addition sign to make an addition calculation (for example 8 + 4). Give the pupil on the right of the pupil a number between 10 and 20 (for example 14).

- Ask pupils what they notice about the values either side of the calculation, ask: *What is the same and what is different?* Ask if there should be an =, < or > in the calculation and to explain why. In pairs, pupils rehearse a justification based on their instant recall of number facts to say to the class. For example:

- Pupils should be able to explain that 14 must be greater than 8 + 4 because they know that 10 + 4 is equal to 14 and 10 is greater than 8. Repeat with further calculations until pupils become more fluent in drawing on prior knowledge of addition facts to justify the use of a <, > and = symbol.

- Expand the task to include two calculations:

- 9 is greater than 8, so 9 added to 4 must be greater than 8 added to 4. Repeat with further calculations until pupils become more fluent in drawing on prior knowledge of addition facts to justify the use of a particular symbol.

- Pupils should then complete Practice Book Question 4.

  Look out for … pupils who continue to calculate the answer to each calculation rather than use what they know about the numbers and their instant recall of number facts.

Chapter 3 Numbers up to 20 and their addition and subtraction   Unit 3.6 Practice Book 1A, pages 91–93

Same-day intervention

- Pupils should be given the opportunity to develop their understanding of commutativity and how the total remains unchanged if the order of the addends are changed. Use cubes to lay out two groups of numbers to represent 8 + 5. Lay out a second two groups to represent 5 + 8. Calculate the totals of each and then place the cubes for 8 + 5 into one side of a balance scales and the cubes for 5 + 8 in the other. Point out that the two pans balance and this is because 8 + 5 is equal to 5 + 8.

Same-day enrichment

- Give pupils a set of calculations with one of the numbers as the unknowns rather than the =, < or > symbol, for example 8 + ? < 7 + 6. For each calculation, work systematically to find a range of numbers that would make the calculation correct.

Challenge and extension question

Question 5

5 Write the missing numbers.

☐ – 3 = 7   ☐ – 6 = 9   ☐ – 9 = 5   ☐ – 8 = 4

☐ – 7 = 5   ☐ – 9 = 8   ☐ – 6 = 6   ☐ – 7 = 9

The challenge and extension question provides pupils with the opportunity to apply what they have learned in this unit. The questions require them to apply their knowledge of equivalence, the relationships between addition and subtraction and known facts to solve problems where the unknown value is the minuend rather than the difference.

Chapter 3 Numbers up to 20 and their addition and subtraction

Unit 3.7
Addition and subtraction (II) (3)

Conceptual context

The previous units have developed pupils' conceptual understanding of the structure of numbers and how flexible partitioning and bridging can be effective calculation strategies for addition. In this unit, pupils' understanding of flexible partitioning is applied to subtraction calculations that focus on counting back and bridging. This approach develops pupils' fluency with partitioning as well as deepening their conceptual understanding of subtraction as the inverse of addition. In order to make explicit the connections between addition and subtraction, similar models, resources and images are used to help make those connections explicit. Some of the questions from Unit 3.5 are echoed here.

Learning pupils will have achieved at the end of the unit

- Pupils will have become more fluent in counting backwards (Q1)
- Pupils will have reinforced their understanding of partitioning numbers into 10 and some more and will have applied this knowledge to subtraction questions (Q1, Q2)
- Pupils will have developed their understanding of the inverse relationship between addition and subtraction (Q2)
- Pupils will have developed their understanding of bridging and how this applies to subtraction questions (Q2, Q3, Q4)
- Pupils will have become more fluent with using symbolic notation in a variety of ways (Q1, Q2, Q3, Q4)
- Pupils will have extended their understanding of part–whole relationships and will have applied this to subtraction (Q1)

Resources

ten-frames; counters; coloured number rods; number lines; coloured flat number tiles; mathematical balance; bar models; mini whiteboards

Vocabulary

partition, bridging, subtract, count back, addend, subtrahend, minuend, part, whole

Chapter 3 Numbers up to 20 and their addition and subtraction   Unit 3.7 Practice Book 1A, pages 94–96

Question 1

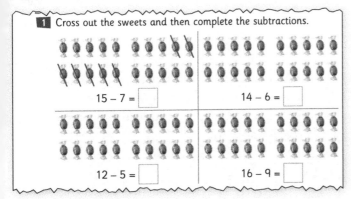

What learning will pupils have achieved at the conclusion of Question 1?

- Pupils will have become more fluent in counting backwards.
- Pupils will have reinforced their understanding of 10 and some more and will have applied this knowledge to subtraction questions.
- Pupils will have become more fluent with using symbolic notation in a variety of ways.
- Pupils will have extended their understanding of part–whole relationships and apply this to subtraction

Activities for whole-class instruction

- Counting back, number songs. In order to subtract efficiently, pupils must be fluent with counting backwards. They should regularly sing songs such as 'Ten (or 20) green bottles', 'There were ten in the bed', 'Five little men in a flying saucer' and so on. Initially, ensure that a number line is used to help pupils keep track of the numbers they have counted and to see the next number they will say.
- 10 and some more revisited. Use a ten-frame and lay out 13 counters. Use subitising to find each addend and write 7 + 6 = 13. Reinforce the language of addition: addend + addend = sum. Remind pupils and demonstrate that, when adding, the two addends are the two parts that are combined to make the whole. Ask pupils how many are in the whole and how many are in each part.

- Ask pupils how they would calculate 13 subtract 6 using the ten-frame and model removing (concrete) and crossing out (pictorial) each one of the 6 as you count backwards.

- Ensure that pupils recognise that the answer to the subtraction calculation is the number of counters left once the subtrahend has been removed.
- Pupils should complete Question 1 in the Practice Book.

Same-day intervention

- Help pupils develop fluency with counting backwards. Using a number line as a visual aid, ask pupils to tell you (i) the number that is 1 less, 2 less or 3 less than a number you say and (ii) say that the number 8 is 1 less than a number you are thinking of. Ask pupils: *What number am I thinking of?*

Same-day enrichment

- If I know ... what else do I know? Provide a few subtraction questions. Pupils should identify the part and whole and use this to write associated addition facts. They may provide two addition questions reflecting their understanding of commutativity and addition.

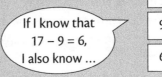

Question 2

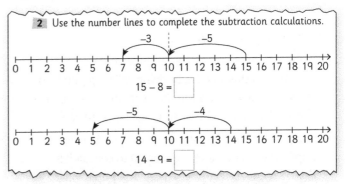

What learning will pupils have achieved at the conclusion of Question 2?

- Pupils will have developed their understanding of bridging and how this applies to subtraction questions.
- Pupils will have developed their use of number lines to include solving subtraction questions.
- Pupils will have become more fluent with using symbolic notation in a variety of ways.

Chapter 3 Numbers up to 20 and their addition and subtraction

Unit 3.7 Practice Book 1A, pages 94–96

Activities for whole-class instruction

Flexible partitioning

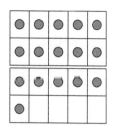

- Using the ten-frame, lay out 16 counters and use subitising to help work out the number of counters. Write the addition sentence 10 + 6 = 16.
- Ask pupils if they can think of two ways to solve the calculation 16 − 8. They could (i) count back in ones from 16 until 8 have been subtracted, or (ii) partition the 8 into 6 + 2, subtract 6 and then 2. Use a number line and a ten-frame to help illustrate the steps taken. In both cases, show counting back in ones and then bridging (−6 to get to 10, then −2). Emphasise that bridging, using known facts to partition, is more efficient than counting back in ones. Discuss whether pupils prefer a number line or a ten-frame. Agree that both are good at different times and that some people might prefer one or the other.
- Pupils should complete Question 2 in the Practice Book.

 ... pupils who rely on counting back in ones for all subtraction calculations.

Same-day intervention

- Using a number line to model subtraction calculations that involve bridging will help pupils to recognise how close a number is to 10 and to use this knowledge to make their strategy more efficient.
- Using a 0–20 number line, mark the number 15. Ask pupils how many more it is than 10 to consolidate that 15 is equal to 10 and 5 more. Ask pupils what 15 subtract 5 would equal and model this on the number line. Then model the calculation 15 − 6. Ask pupils how they could do this, emphasising that 6 is 1 more than 5 and model the steps on the number line.
- Repeat with other pairs of subtraction calculations that emphasise how the subtrahend in the second calculation can be partitioned.

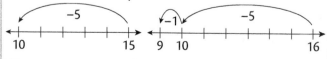

Same-day enrichment

- Ask 'empty box' questions that focus on the partitioning of the subtrahend. These can be presented as written calculations, on a ten-frame or number line. In each case, pupils should identify which number had been partitioned to aid bridging and how it has been partitioned. Remind pupils = means 'has the same value as' or 'is equivalent to'.

  14 − ☐ = 14 − ☐ − 3 = 7   17 − 9 = 17 − 7 − ☐ =
  13 − 8 = 13 − ☐ − 5 = ☐

Question 3

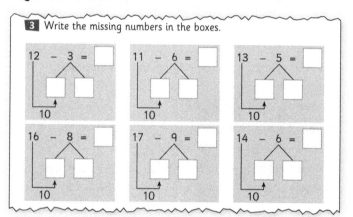

What learning will pupils have achieved at the conclusion of Question 3?

- Pupils will have developed their understanding of bridging and how this applies to subtraction questions.
- Pupils will have become more fluent with using symbolic notation in the context of subtraction.

Activities for whole-class instruction

- In pairs, give pupils sets of 1–9 digit cards. Write on the board the minuend (a number between 11 and 20) and the subtrahend (a number that would require crossing the 10). Each pair selects the appropriate digit cards to show how the subtrahend is partitioned to reach 10.
- Provide subtraction questions that can be recorded first using a number line to highlight how the subtrahend has been partitioned to reach 10. Develop this into a bridging diagram (below) and highlight the similarities between the two written representations while providing opportunities for pupils to become fluent in the language of partitioning the subtrahend when bridging.

Chapter 3 Numbers up to 20 and their addition and subtraction

Unit 3.7 Practice Book 1A, pages 94–96

- Give the class a pair of numbers to subtract. Ask them to find the answer using a number line. Re-introduce pupils to the bridging diagram from Unit 3.5 and ask them to draw the number line that matches the bridging diagram. Each time, rehearse orally the steps that they are going through. Ask pupils: *What is the same? What is different?* about each representation. Emphasise that they both show the two steps involved in bridging: partition to reach 10 and then counting back.

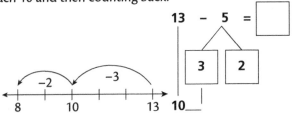

- Pupils should then complete Question 3 in the Practice Book.

 Partition the subtrahend to reach 10, then count back.

 ... pupils who cannot move fluently between the different representations (whose understanding appears to be procedural rather than conceptual).

Same-day intervention

- Pupils should use a mathematical balance to develop fluency with number bonds. This will help them when partitioning the subtrahend to bridge through 10 when subtracting. Challenge them to find all the number bonds (or addition sentences) for all the numbers from 4 to 10, starting with 4 and only going to 5 when they have found all those for 4. They should find the patterns in their answers.

- Pupils should use a number line to show how to work out 13 – 8 = 7. Check that they are using mathematical language correctly. Emphasise the two-step nature of the bridging strategy: reach a 10 and then count back some more. Can pupils see the connection between the work with the mathematical balance and this activity? (Knowing the number bonds helps them know how to partition the subtrahend.)

- Draw the two-step representation and support pupils to identify the two steps and use appropriate mathematical language to describe them. Ask pupils where the first step is shown on the number line.

Same-day enrichment

- Play 'Four-in-a-row' to develop fluency with addition and subtraction.
- The aim of the game is to cross out four consecutive numbers on a number line from 0 to 20. Pupil 1 selects one number from Box A and one from Box B. They decide whether these numbers will be added or subtracted, calculate the answer and write their initial on that number on the number line. Pupil 2 then repeats the process but cannot make a calculation that has the same answer as has been made previously.
- The game continues until one pupil has written their initials on four consecutive numbers.

Question 4

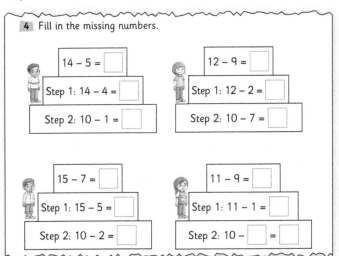

What learning will pupils have achieved at the conclusion of Question 4?

- Pupils will have developed their understanding of bridging and how this applies to subtraction questions.
- Pupils will have become more fluent with using symbolic notation in the context of subtraction.

Activities for whole-class instruction

- Pupils have rehearsed the bridging strategy and should have become fluent in applying this strategy to calculations with numbers up to 20. They have also already used a variety of recording representations: ten-frames, number lines and bridging diagrams. The aim now is to increase fluency with bridging and using written recording as a means of recording the steps rather than as an aid to help identify them.

Chapter 3 Numbers up to 20 and their addition and subtraction

Unit 3.7 Practice Book 1A, pages 94–96

- Step-by-step recording. As the pupil explains the strategy, you record it, first using the number line, then the bridging diagram and then using the step model.

  15 – 8 =
  15 – 5 = 10
  10 – 3 = 7

- Pupils then complete Question 4 in the Practice Book.
- Ask pupils what is the same and what is different about the forms of recording. Clarify that they each show the steps involved in bridging but that the step model is more efficient.
- Repeat with further calculations.

   … *Partition the subtrahend to reach 10, then count back.*

- Give pupils the whole and one of the parts and ask them how they would work out the missing part, identify how it would be partitioned and use addition to help justify their solution. For example: 14 – ☐ = 6:

14	
6	?

  14 subtract something is equal to 6. I know that 6 added to 8 is equal to 14, so the missing part must be 8. If I didn't know it, I could work it out by counting up from 6 to 14.

  Or I could subtract 6 from 14. I would partition the 6 into 4 and 2 to reach 10 and then count back 2 more.

Same-day intervention

- Some pupils need to experience a concept for longer than others before it becomes fully understood and connected to other concepts that they know. Therefore, spend time using the number lines again to help solve subtraction calculations using bridging. Ensure that pupils are using mathematical language correctly and are able to describe and demonstrate how they have partitioned the subtrahend to reach 10. Once their explanations are fluent, then begin to use the step-by-step model alongside the number line. As each step on the number line is completed, complete the next step in the step-by-step diagram to help pupils see how the two representations are different ways of presenting the same strategy, using the same numbers and the same language.

Same-day enrichment

- Working backwards and forwards. In order to develop fluency with partitioning and deepen understanding of bridging, provide pupils with some missing number questions. This relationship between known and unknown numbers can be modelled in a variety of ways. In Chapter 2 pupils used bar models – it may be helpful to begin by using bar models again to help pupils to develop connections between the mathematics they have learned in different units.

Challenge and extension question

Question 5

5 Fill in the ◯ so the sum of the numbers on each line is 20.

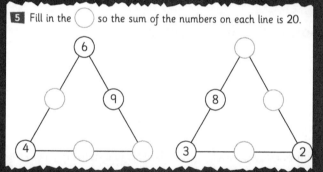

This problem develops fluency with addition and subtraction of single-digit numbers. The presentation of known and unknown numbers in a pyramid encourages pupils to select counting on or counting back strategies or to apply their knowledge of number facts to reach the solution. A successful strategy will involve identifying the most efficient starting point (where two values are known) and then identifying whether addition or subtraction is required to find the other missing values.

Chapter 3 Numbers up to 20 and their addition and subtraction

Unit 3.8
Addition and subtraction (II) (4)

Conceptual context

So far in the chapter, pupils have developed their understanding of the structure of numbers between 10 and 20, applied their understanding of the commutative law (when re-ordering) and partitioned and used bridging as strategies for solving addition and subtraction questions. It is important that pupils maintain and develop fluency and flexibility with calculation strategies and are able to select the most efficient one. Efficiency is further developed by encouraging pupils to count on and back rather than count all and to continue to subitise to work out how many objects are in a group. It is also important that pupils continue to develop and apply their understanding of the relationship between addition and subtraction and when they may use their knowledge of one operation to help solve a question involving the other. To this end, this unit continues to make explicit connections between the operations of addition and subtraction through the use of concrete resources, images and symbols.

Learning pupils will have achieved at the end of the unit

- Pupils will have reinforced the connections between symbolic and pictorial representations of calculations (Q1)
- Pupils will have practised a range of calculation strategies including counting back, counting on and bridging and used them appropriately (Q1, Q2, Q3)
- Pupils will have continued to develop their use of mathematical terms such as addend, sum, minuend, subtrahend, difference, part and whole (Q1, Q2, Q3)
- Pupils will have extended their experience of using particular strategies and will be able to explain when a strategy is most appropriate (Q1, Q2, Q3, Q4)
- Pupils will have developed their understanding of the relationship between addition and subtraction (Q3)
- Pupils will have consolidated their instant recall of addition and subtraction facts and used them to solve problems (Q4)

Resources

buttons; bowls; boxes with lids; coloured flat number tiles; number lines; counters; ten-frames; double-sided counters; objects such as apples; mini whiteboards

Vocabulary

part–part–whole, addend, subtrahend, minuend, total, difference, partitioning, bridging

Chapter 3 Numbers up to 20 and their addition and subtraction Unit 3.8 Practice Book 1A, pages 97–99

Question 1

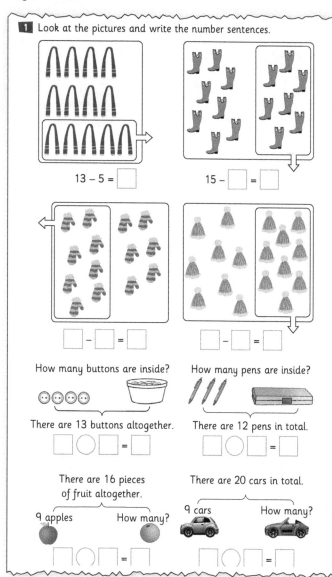

What learning will pupils have achieved at the conclusion of Question 1?

- Pupils will have reinforced the connections between symbolic and pictorial representations of calculations.
- Pupils will have practised a range of calculation strategies including counting back, counting on and bridging and used them appropriately.
- Pupils will have continued to develop their use of mathematical terms such as addend, sum, minuend, subtrahend, difference, part and whole.
- Pupils will have extended their experience of using particular strategies and will be able to explain when a strategy is most appropriate.

Activities for whole-class instruction

- Part–part–whole. Have two small bowls each containing a handful of buttons. Explain that there are two parts. Ask pupils: *If these are the parts, how will I find out how many buttons are in the whole?*
- Count the number of buttons in each bowl and clarify that each part is an addend and that the number in the whole gives us the sum.
- Display the button groups and the addition number sentence. Ask pupils: *How many parts are there? How many are in the part? How many are in this part? What is the total amount of buttons in the whole?*
- Ask pupils to consider, if they know that 6 + 7 = 13, what else do they know?
- Ensure that all of the associated facts are mentioned: 6 + 7 = 13 (because addition is commutative), 13 − 7 = 6 and 13 − 6 = 7.

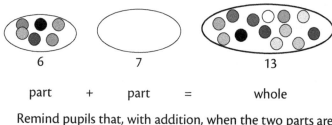

part　　　+　　　part　　　=　　　whole

Remind pupils that, with addition, when the two parts are added you find the sum. With subtraction you know the whole and you are working out the size of the remaining part once one part has been removed. There are 13 buttons in the whole, when 6 are subtracted 7 are left.

- Repeat the process with one group of buttons placed in a tub with a lid.

- Ask pupils a series of questions that require them to use their knowledge of number facts and their understanding of the inverse relationship between addition and subtraction.
  - If there are 8 buttons in the box pictured, and 6 more buttons, how many buttons are there altogether?
  - If there are 11 buttons in the box, how many are there altogether?
  - If there are 18 buttons altogether, how many are in the box? This question can be written as 18 − 6 = ? or 6 + ? = 18, and solved either by counting up from 6 to 18, or back 6 from 18.

Chapter 3 Numbers up to 20 and their addition and subtraction

Unit 3.8 Practice Book 1A, pages 97–99

- The whole, the total number of buttons in the group, is 18 and one part (the number of buttons outside the box) is also known. The question requires pupils to find the value of the other part, which is the number of buttons in the box. Use the number line to model the calculation and help pupils to identify that either counting on or counting back can be used to find the answer.

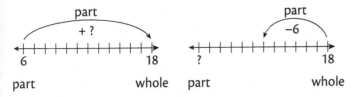

- Develop pupils' fluency with identifying both strategies by asking similar questions such as: *If there are 12 buttons altogether, how many are in the box?*
- Pupils should complete Question 1 in the Practice Book.

 When two parts are added we find the whole; when one part is subtracted from the whole, we find the value of the other part.

Same-day intervention

- In order to become fluent in bridging when subtracting, some pupils may need to continue to use manipulatives to model how the subtrahend is partitioned. Ten-frames and double-sided counters can be used to model this process. For example, the solution to the calculation 16 – 8 can be modelled by representing the 16 as 10 and 6 more in the ten-frame. The subtrahend is then partitioned into 6 and 2, which is shown by showing the other side of the counters.
- With ten-frames, pupils can easily see how the subtrahend should be partitioned as the counters sit in the different section of the frame. Rehearse with other calculations involving bridging until pupils are able to describe the process fluently and recognise how the subtrahend should be partitioned in each case to get back to 10.

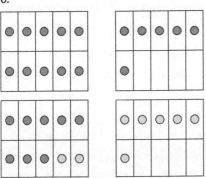

Same-day enrichment

- In order to extend pupils' understanding of part–part–whole relationships and the inverse relationship between addition and subtraction, pupils will now work with three parts that total a whole, for example 6 + 7 + ? = 19. Pupils should be able to identify how many parts there are in the whole, and how they would work out the value of the missing part. This could be through a combination of addition and subtraction calculations. Pupils should identify whether they always use the same method or whether some combinations of numbers prompt them to use other strategies. For example, with the calculation ? + 4 + 7 = 14, pupils may notice that the whole is equal to 14, one part is 4, so the other two parts must equal 10.

Question 2

2 Complete the table.

minuend	11	13	15
subtrahend	5	4	7
difference			

minuend	16	12	14
subtrahend	8	9	6
difference			

What learning will pupils have achieved at the conclusion of Question 2?

- Pupils will have practised a range of calculation strategies including counting back, counting on and bridging.
- Pupils will have continued to develop their use of mathematical terms such as addend, sum, minuend, subtrahend and difference, and part and whole.
- Pupils will have extended their experience of using particular strategies and will be able to explain when a strategy is most appropriate.

Activities for whole-class instruction

- Pupils saw calculations presented in this form in Unit 3. When the operation was addition, the focus was on deciding when it was appropriate to re-order the numbers to increase efficiency. It is important that pupils understand that the law of commutativity does not apply to subtraction and instead, the focus here is on developing

Chapter 3 Numbers up to 20 and their addition and subtraction Unit 3.8 Practice Book 1A, pages 97–99

pupils' fluency with the language of subtraction. As has been the case previously, the numbers in the calculations are those that pupils will be familiar with, in order for them to focus on their correct use of language.

- Lay out a quantity of small objects (for example 12) so that they are easy to subitise.
- Point to the whole and check that pupils remember that in subtraction it is called the minuend. Demonstrate removing a quantity of the objects, say four. Remind pupils that the part we subtract is called the subtrahend.

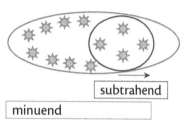

- Ask further questions that require pupils to use their knowledge of the relationship between minuend, subtrahend and difference. For example: *If the minuend is 16 and the difference is 9, what was the subtrahend?* and *If the difference is 5 and the subtrahend is 7, what was the minuend?* Encourage pupils to explain their solutions using manipulatives to support their explanations as appropriate.
- Pupils should complete Question 2 in the Practice Book.

 The whole is the minuend, the part we subtract is the subtrahend; the result of subtracting one number from another is called the difference.

Same-day intervention

- Developing fluency with mathematical language is important in developing conceptual and procedural understanding of mathematics. For some pupils it takes longer for the words to have meaning and for them to be able to associate the words with particular examples.
- Use ten-frames to create addition and subtraction sentences, where the whole is between 11 and 20 and one part is 10 or more. First model an addition sentence and rehearse the language of part–part–whole.
- Once pupils have become more fluent with identifying the parts and whole with addition, move on to subtraction and then increase difficulty by introducing bridging through 10. Focus on understanding of parts and wholes and knowing that, when two of the three elements are known, the other can be deduced.

Same-day enrichment

- What could the parts be? This task requires pupils to work systematically to identify all possible parts that could make a whole and to deduce which part is hidden within each box.
- Start with two boxes with lids. Secretly place some (for example five) small identical counting objects in one box and put on the lid and some (for example seven) in the other box and put on the lid.
- Tell pupils that the whole is 12 and ask them what the parts could be. Encourage pupils to work systematically to ensure that they have every possible combination.
- Once they have written their list, they can pick up the boxes, compare them and shake them. They can use this to help eliminate parts that cannot be possible (such as 1 + 11, 2 + 10). Pupils cross out all of the impossible answers and then discuss which they think is most likely from the remaining possible ones. They suggest a quantity for one box and the lid is removed to reveal the numbers of objects within.
- Pupils then work out the quantity of the other part as the remaining part will be the difference between the exposed part and the whole.

Question 3

3 Complete the subtraction calculations.

11 − 2 = ☐	11 − 8 = ☐	12 − 5 = ☐	13 − 8 = ☐
12 − 3 = ☐	12 − 4 = ☐	14 − 6 = ☐	16 − 9 = ☐
13 − 4 = ☐	12 − 6 = ☐	15 − 7 = ☐	13 − 6 = ☐
14 − 5 = ☐	14 − 7 = ☐	12 − 8 = ☐	12 − 7 = ☐

What learning will pupils have achieved at the conclusion of Question 3?

- Pupils will have practised a range of calculation strategies including counting back, counting on and bridging.
- Pupils will have continued to develop their use of mathematical terms such as addend, sum, minuend, subtrahend, difference, part and whole.
- Pupils will have extended their experience of using particular strategies and will be able to explain when a strategy is most appropriate.

Chapter 3 Numbers up to 20 and their addition and subtraction

Unit 3.8 Practice Book 1A, pages 97–99

Activities for whole-class instruction

- Building on pupils' instant recall of subtraction facts, this activity ensures that their knowledge of them is not just rote, but that they can use known facts to help answer related facts.
- Start with one subtraction question, for example 15 − 7 = 8. Focusing only on subtraction, ask pupils to work in pairs to suggest other subtraction questions they could answer by adapting this fact. Expect pupils to be able to suggest a subtraction calculation as well as provide an explanation of how they used the known fact. For example, pupils might say: *I know that 15 − 7 = 8, so 15 − 8 = 7. I have made the subtrahend one bigger so the difference will be one smaller.*
- What size steps? Pupils need to be able to partition the subtrahend with fluency so should be able to identify how the number of ones in the minuend determines how the subtrahend is partitioned.
- Provide a series of calculations with the same minuend, for example 11 − 3, 11 − 4, 11 − 5, 11 − 6, 11 − 7. Clarify that when subtracting from 11, the subtrahend will be partitioned into 1 and some more because 11 is 1 more than 10.
- Use a number line to model the steps, each time re-affirming the size of the first partition required when bridging.
- Repeat this with other minuends clarifying that the size of the first step should always reach 10.
- Pupils should complete Question 3 in the Practice Book.

  All say... *Partition the subtrahend to reach 10, then count back.*

  Look out for ... pupils who are counting back using their fingers and are unsure whether the answer is the last number they say, or the number that remains.

Same-day intervention

- Use a set of eight objects such as apples. Explain that you are going to solve the problem 8 − 3. Enact the problem explaining that these three apples form one part of the whole.
- Explain that the whole, the minuend, is equal to 8; the part we are subtracting, the subtrahend, is 3. The other part is the number of apples that remain once we have subtracted 3, so they need to know the quantity that are left, (and not the last number they may have said out loud when counting backwards).
- Once pupils have become fluent with this as a concrete activity, move on to a number line so that pupils can see that the size of the second part is the number they count back to.
- If time permits, move on to teens numbers. Provide pupils with questions with carefully chosen numbers so that they can recognise patterns that occur when subtracting from teens minuends and consolidate their understanding of how the minuend helps to identify how to partition the subtrahend.

13 − 3 =	14 − 4 =	15 − 5 =
13 − 4 =	14 − 5 =	15 − 6 =
13 − 5 =	14 − 6 =	15 − 7 =
13 − 6 =	14 − 7 =	15 − 8 =

Same-day enrichment

- In order to ensure that pupils develop fluency with calculation strategies and develop their understanding of the relationship between addition and subtraction, the focus is on how you could find the minuend and the subtrahend if only the difference was known (☐ − ☐ = 4).
- In pairs, pupils suggest one minuend and subtrahend that has a difference of 4 (12 subtract 8 is equal to 4) and their partner checks using the corresponding addition question (8 added to 4 is equal to 12), or other known facts (8 is two lots of 4, 12 is three lots of 4, so the difference between 12 and 8 must be one lot of 4).

Chapter 3 Numbers up to 20 and their addition and subtraction   Unit 3.8 Practice Book 1A, pages 97–99

Question 4

4 Write >, < or = into each ◯.
13 – 8 ◯ 7   14 – 9 ◯ 7   6 ◯ 13 – 7   9 ◯ 14 – 6
14 – 5 ◯ 8   16 – 9 ◯ 5   10 ◯ 18 – 9   7 ◯ 11 – 9

What learning will pupils have achieved at the conclusion of Question 4?

- Pupils will have extended their experience of using particular strategies and will be able to explain justify use of a particular strategy.
- Pupils will have consolidated their instant recall of addition and subtraction facts and used them to solve problems.

Activities for whole-class instruction

- Revisit the 'Get the balance right' activity from Unit 3.6. This time, the pupil in the middle has a =, < and > symbol that can be used.
- Begin with quick-fire questions involving subtraction with numbers less than 20. Vary the language of the question to include *x take away y, subtract y from x, what is x subtract y? If I subtract xx from y, what is the difference?*
- Then, starting with a minuend and a subtrahend and a difference, ask pupils what they notice about the values either side of the calculation. *What is the same and what is different?* Ask if there should be an =, < or > in the calculation and how they know.
- In pairs, pupils rehearse a justification that is based on their instant recall of number facts to say to the class. For example:

- Pupils should be able to explain that 11 – 6 is less than 6 because:
  (i) *I know that 6 added to 5 is equal to 11, so the difference between 11 and 6 is 5, which is less than 6*
  (ii) *double 6 is 12 and as the minuend is smaller than 12, the difference must be less than 6.*

Use a resource such as flat coloured number tiles and number lines to model the calculation and to support the pupil's explanations.

- Pupils should complete Question 4 in the Practice Book.

Same-day intervention

- Rehearse strategies that focus on the numbers in a calculation and relationships between them, rather than on calculating the answer, such as: *If I know this, what else do I know?*
- Starting with a simple addition calculation, such as 13 – 8, pupils identify the facts they can instantly recall and those that they can quickly derive from this fact. Resources such as counters and number lines can help pupils to make connections between known and quickly derived facts

  8 + 5 = 13
  If 8 and 5 are the parts, then 13 is the whole.

Same-day enrichment

- Expand the task to include two calculations.

- This will not only develop pupils' instant recall but also help them to see the relationships between facts.

Challenge and extension question

Question 5

5 Look at the numbers first and then fill in the boxes.
8 – 1 = 7   ☐ – ☐ = 7   ☐ – ☐ = 7
9 – 2 = 7   ☐ – ☐ = 7   ☐ – ☐ = 7
10 – 3 = 7   ☐ – ☐ = 7   ☐ – ☐ = 7
☐ – ☐ = 7   ☐ – ☐ = 7   ☐ – ☐ = 7

This problem provides pupils with the opportunity to develop instant recall of subtraction calculations with a given difference; also to think flexibly about the number facts that they know as they start with the difference, not the minuend and subtrahend. It also provides an opportunity for pupils to recognise the importance of working systematically in order to find all of the possibilities.

Chapter 3 Numbers up to 20 and their addition and subtraction

Unit 3.9
Addition and subtraction (II) (5)

Conceptual context

Pupils have developed a range of strategies for adding and subtracting: use of known or quickly derived facts, bridging and re-ordering. They have also developed their understanding of the relationships between addition and subtraction, the nature of the parts and the whole and how this can be used to help solve problems. It is important that pupils are confident with a range of strategies and maintain flexibility when calculating so that they can select the most appropriate strategy for the problem they are solving. It is also important that they are fluent in the use of appropriate mathematical language, and can use and interpret symbols.

This unit therefore provides questions that can be solved using either addition or subtraction, questions that encourage efficient calculating (counting on rather than counting all) as well as presenting questions in a range of representations.

Learning pupils will have achieved at the end of the unit

- Pupils will have developed fluency in counting on (rather than counting all) (Q1)
- Pupils will have applied their understanding of commutativity (re-ordering) and bridging to a range of calculations (Q1, Q2)
- Pupils will have developed their knowledge of 0 and have investigated the effect it has on addition and subtraction situations (Q2)
- Pupils will have had further opportunities to work with a range of representations including images, numerals and diagrams (Q1, Q2, Q3, Q4)
- Pupils will have continued to develop a range of calculation strategies that they can apply appropriately and efficiently (Q1, Q2, Q3, Q4)
- Pupils will have extended and applied their understanding of part–whole relationships to addition and subtraction situations (Q1, Q2, Q3, Q4)
- Pupils will have developed greater fluency in using the terms addend, total, minuend, subtrahend, difference and part–whole (Q1, Q2, Q3, Q4)
- Pupils will have had further opportunities to develop rapid recall of number facts (Q2, Q3)
- Pupils will have investigated the relationships between numbers in calculations and used this to solve a range of missing number problems (Q2, Q3, Q4)

Resources

mini whiteboards; number fans or digit cards; coloured number rods; coloured strips of card; interlocking cubes; counting resources (e.g. buttons, shells and so on); 0–20 number lines; bar models

Vocabulary

total, addend, part–part–whole, minuend, subtrahend, difference

Chapter 3 Numbers up to 20 and their addition and subtraction   Unit 3.9 Practice Book 1A, pages 100–102

Question 1

What learning will pupils have achieved at the conclusion of Question 1?

- Pupils will have had further opportunities to work with a range of representations including images, numerals and diagrams.
- Pupils will have developed fluency in counting on (rather than counting all).
- Pupils will have applied their understanding of commutativity (re-ordering) and bridging to a range of calculations.
- Pupils will have continued to develop a range of calculation strategies that they can apply appropriately and efficiently.
- Pupils will have extended and applied their understanding of part–whole relationships to addition and subtraction situations.
- Pupils will have developed greater fluency in using the terms addend, total, minuend, subtrahend, difference and part–whole.

Activities for whole-class instruction

- Whole-class recall of addition and subtraction facts. Give each pupil a number fan or set of digit cards, which they can spread around in front of them.
- Ask an addition question such as 8 + 6. Count down 3, 2, 1 (quickly) and then say *Show me*. Pupils use the number fan or cards to show you the answer. Increase the speed of the 3, 2, 1 as necessary to ensure that pupils are using their knowledge of addition and subtraction facts and do not have time to use their fingers to calculate.
- For each question, vary the way in which the question is phrased. For example *What is the total of 12 and 7? What do I add to 5 to equal 14? What is 8 and 6 altogether?*
- *Part–part–whole; what do I know* (1)? Start by using counters and lay out two groups, for example a group of 5 and a group of 7. Ask pupils how many parts there are (there are two parts) and what is the whole. Ask them how they worked out the whole. They should explain that they added the parts (addends).
- Write the calculation up for pupils to see and write part–part–whole, addend and total underneath. Emphasise that when the whole or sum was unknown, the parts or addends were added. Rehearse with further questions where the sum is unknown if necessary.
- Then ask: *What happens if I know the sum, but not one of the parts?* For example ☐ + 5 = 14. Remind pupils that they learned they could use bar models to help them see what to do in this situation.
- There is another way to find out the value of the unknown part. Attach three strips of card to the board, replicating the bar model above, that is, two parts of different sizes and different colours and a corresponding whole in a third colour.
- Ask for volunteers to demonstrate the other way to find the value of the missing part. Can pupils show that removing the known part leaves the unknown part, so subtracting the value of the part that is known from the whole will reveal the value of the part that is unknown? Repeat several times, talking about 'taking away' the known part as you physically 'take it away' from the bar model.
- *Part–part–whole, what do I know* (2)? Show a similar bar model:

a	
b	c

- Say: *If the whole is 20 and part b is 15, what is the value of part c?*

Chapter 3 Numbers up to 20 and their addition and subtraction

Unit 3.9 Practice Book 1A, pages 100–102

Ensure pupils can explain that to find the difference, they must subtract the subtrahend (one of the parts) from the minuend (the whole). Write the calculation up for pupils to see and write part–part–whole, minuend, subtrahend and difference underneath.

All say... When adding, I can find the whole by adding the parts. If I know the whole, I can find one part by subtracting the other part from the whole.

- Then ask: *What happens if I know the difference, but not the whole?* For example ☐ − 5 = 14. Ensure that pupils know that the whole (minuend) is unknown. Give pupils a mini version of the bar model:

- Challenge them to find the whole (the minuend) when ☐ − 5 = 14. Pupils should work in pairs and rehearse their explanation of their strategy for finding the minuend. (*The minuend is the whole, and if I know the whole, I can find one part by subtracting the other part from the whole.*) Rehearse with further questions where the minuend is unknown if necessary.

- Pupils should complete Question 1 in the Practice Book.

Same-day intervention

- Use coloured number rods to revise the language of part–part–whole, addend and total, minuend, subtrahend and difference.
- Rehearse the language alongside enacting the operation of part added to part is equal to the whole, addend added to the addend is equal to the sum. The order in which the parts are added doesn't change the whole as addition is commutative. Explain this in terms of words, then in terms of numbers. (*If 5 and 3 are the parts, 8 is the whole. 5 added to 3 is equal to 8, 3 added to 5 is equal to 8.*)

- When subtracting, the order in which the parts are used changes what they are called. Whichever part is subtracted is called the subtrahend, whichever part remains is called the difference.

whole/sum
| part/addend | part/addend |

- Explain this in term of words, then in terms of numbers. If the whole or minuend is 13, and one part, equal to 5, is subtracted (the subtrahend) the difference is equal to 8. If the whole or minuend is 13, and one part, equal to 8, is subtracted (the subtrahend) the difference is equal to 5.

whole/minuend
| part | part/addend |

whole/minuend
| subtrahend | difference |

whole/minuend
| subtrahend | difference |

Same-day enrichment

- Put the following number sentences into a context:

6 + 5 = 11   5 + 6 = 11
11 − 6 = 5   11 − 5 = 6
☐ + 5 = 11   6 + 5 = 11
11 − ☐ = 5   ☐ − 6 = 5

Question 2

2. Complete the following calculations.

12 − 0 =   5 + 7 =   15 − 8 =   13 − 5 =

9 + 6 =   11 − 3 =   9 + 0 =   12 − 4 =

7 + 6 =   14 − 7 =   13 − 8 =   19 + 1 =

What learning will pupils have achieved at the conclusion of Question 2?

- Pupils will have had further opportunities to work with a range of representations including images, numerals and diagrams.
- Pupils will have applied their understanding of commutativity (re-ordering) and bridging to a range of calculations.
- Pupils will have continued to develop a range of calculation strategies that they can apply appropriately and efficiently.
- Pupils will have extended and applied their understanding of part–whole relationships to addition and subtraction situations.
- Pupils will have developed greater fluency in using the terms addend, total, minuend, subtrahend, difference and part–whole.

Chapter 3 Numbers up to 20 and their addition and subtraction   Unit 3.9 Practice Book 1A, pages 100–102

- Pupils will have developed their knowledge of 0 and have investigated the effect it has on addition and subtraction situations.
- Pupils will have had further opportunities to develop rapid recall of number facts.
- Pupils will have investigated the relationships between numbers in calculations and used this to solve a range of missing number problems.

Activities for whole-class instruction

- Although bridging remains the focus for the calculation strategy, pupils are now introduced to questions that involve 0 or 20. This has extended the range of the numbers pupils have worked with and they need to develop their understanding of the effect 0 has in addition and subtraction calculations.
- Begin by presenting pupils with the calculation 8 + 0. Ask pupils how many parts there are and the size of each part. Ask them what the two addends are. Then ask pupils to use coloured number rods to model the calculation and to explain what they noticed. Emphasise that even though there are two parts, if one addend is a 0 the whole is equal to the other part. Repeat with 0 + 12, and further questions as necessary, asking pupils to create this using coloured number rods. Ask them what they notice about the size of the parts and the whole (the size of the addends and the total).
- Once pupils are able to explain fluently that when an addition calculation involves 0, the whole is equal to the other addend, move on to examining a subtraction calculation involving 0.
- Begin by presenting pupils with the calculation 7 − 0. Ask them what the whole is and the value of the subtrahend. Then ask pupils to use coloured number rods to model the calculation and to explain what they have noticed. Emphasise that, although there is a subtrahend, 0, the difference (the size of the part that remains) will be equal to the minuend (the whole). Repeat with 13 − 0 and further questions as necessary to develop fluency with the model and the explanation.
- Pupils should complete Question 2 in the Practice Book.

   Look out for ... pupils who think that zero is 'nothing' rather than part of the set of counting numbers.

... pupils who think that adding always makes the whole larger.

... pupils who think that subtracting always makes the whole smaller.

Same-day intervention

- What do you know? What do you want? Arrange pupils into pairs. Give each pair five objects (such as buttons). Ask them to split the objects between them and say: how many parts there are; the size of each part (three buttons and two buttons); the size of the whole (sum) and write this as a number sentence.
- Ask pupils to find all the different ways they can to have different-sized parts if the sum is 5. If necessary, prompt them to have one arrangement in which one pupil has five objects and the other has zero. Ask pupils to say what the size of each part is and emphasise that the 0 represents no buttons.
- Repeat the task with other objects (such as pencils, shells) and other small numbers smaller than 10. Once pupils have become more fluent with making parts, including 0, and are able to explain what the 0 means in that context, move on to subtraction scenarios.
- From a set of five objects, make the subtrahend 1; what is the difference? Ask: *Do the two parts (subtrahend and difference) equal the whole?* Now from the set of five objects make the subtrahend 2, ask: *What is the difference?* Ask: *Do the two parts (subtrahend and difference) equal the whole?*
- Repeat until the subtrahend is 0 and explain that the 0 means that no objects were subtracted, which means that the minuend (the whole) is equal to the difference. Repeat the task with other objects until pupils are fluent in explaining and demonstrating what happens when you add or subtract 0.

Same-day enrichment

- Developing fluency with using 0. In pairs, pupils take a pack of 0–20 number cards.
- They turn over a card to represent the sum and then make all of the sums they can involving addition and two parts (including a 0), then all of the subtraction questions they can (involving numbers less than 20 including 0). Pupils should be encouraged to explain what they notice about calculating with 0.

Chapter 3 Numbers up to 20 and their addition and subtraction   Unit 3.9 Practice Book 1A, pages 100–102

Question 3

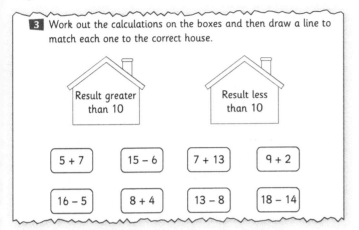

3 Work out the calculations on the boxes and then draw a line to match each one to the correct house.

What learning will pupils have achieved at the conclusion of Question 3?

- Pupils will have had further opportunities to work with a range of representations including images, numerals and diagrams.
- Pupils will have continued to develop a range of calculation strategies that they can apply appropriately and efficiently.
- Pupils will have extended and applied their understanding of part–whole relationships to addition and subtraction situations.
- Pupils will have developed greater fluency in using the terms addend, total, minuend, subtrahend, difference and part–whole.
- Pupils will have investigated the relationships between numbers in calculations and used this to solve a range of missing number problems.
- Pupils will have had further opportunities to develop rapid recall of number facts.

Activities for whole-class instruction

- If I know the sum, what could the addends be? Using a number line from 0 to 20, circle one number (for example 9). Explain that this is the total and that pupils in pairs have to find two numbers to be the addends to make that total.
- For each pair of numbers suggested, draw the steps on the number line and write the number sentences below. Ask pupils what they notice about the size of each addend if the total is below 10 (both addends are less than 10).

- Choose another total (for example 19) and repeat the task. Ask pupils what they notice about the size of each addend if the total is greater than 10 (if one addend is greater than 10, the other addend is less than 10). Test this hypothesis with other totals greater than and less than 10.
- Say: *If I know the difference, what could the minuend and subtrahend be?*

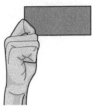

- Explain that the difference is going to be 5. Take a red strip of card to put over the number line to mark the difference (for example between 2 and 7). Ask pupils: *If that difference is 5, what, in this example, are the minuend and the subtrahend?*
- Move the red strip to between 8 and 13, and ask pupils what is the same and what is different about 7 – 2 and 13 – 8 and also 20 – 15 (the difference is 5 but the minuend and subtrahend have changed).
- Ask pupils whether the numbers would be closer together or further apart on the number line if the difference was small? If the difference is always the same and the minuend keeps getting bigger will the subtrahend always get bigger as well? Can pupils explain with some examples?
- Pupils should complete Question 3 in the Practice Book.

Same-day intervention

- Show pupils six interlocking cubes of the same colour. Explain that 6 is the whole and will be the minuend.
- Lay a second row of cubes below the first one and take two cubes away from the top row. Explain that you have taken away two cubes and this is the subtrahend. The part that remains is the difference between the whole you started with and the part you subtracted:

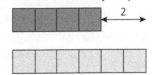

- Provide pupils with interlocking cubes and ask them to lay out two rows of cubes with a difference of 2, difference of 5, difference less than 10.
- For each calculation they make, pupils should be able to state the minuend, the subtrahend and the difference.

Chapter 3 Numbers up to 20 and their addition and subtraction Unit 3.9 Practice Book 1A, pages 100–102

Same-day enrichment

- Say to pupils: *Give me an example of an addition question with an answer of 14, ... and another, ... and another.* Then say: *Give me a really hard question where the answer is 14.* This will encourage adding three parts, or perhaps, combining addition and subtraction.

  Say: *Give me an example of ... a subtraction question with a difference of 6, and another, and another ... really hard question where the answer is 6.*

Question 4

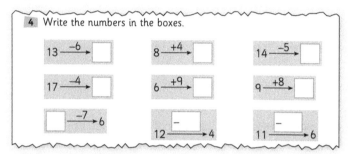

What learning will pupils have achieved at the conclusion of Question 4?

- Pupils will have had further opportunities to work with a range of representations including images, numerals and diagrams.
- Pupils will have continued to develop a range of calculation strategies that they can apply appropriately and efficiently.
- Pupils will have extended and applied their understanding of part–whole relationships to addition and subtraction situations.
- Pupils will have developed greater fluency in using the terms addend, total, minuend, subtrahend, difference and part–whole.
- Pupils will have had further opportunities to develop rapid recall of number facts.
- Pupils will have investigated the relationships between numbers in calculations and used this to solve a range of missing number problems.

Activities for whole-class instruction

- Play 'Guardian of the rule'. Re-introduce the class to a function machine. Explain that when using a function machine a number goes in one side, something happens to it and then a new number comes out from the other side. Pupils have to work out the function of each machine.

  6 → ? → = 9

- Ask pupils what they notice about this function machine. Establish that the number 6 has gone in and the number 9 has come out. If the number that comes out of the function machine is greater than the number that went in, what might be happening in the machine?

- Use coloured number rods to help identify what is known (the first addend and the total) and what is unknown and how the unknown value can be worked out. Number lines will help if needed.

- In pairs, pupils rehearse their explanation of what the function of this machine is and how they know. Explain that it is important to check that they have identified the function correctly by testing it with other numbers.

  6 → ? → = 9

  17 → ? → = 2

- For each question, model using the bar model to show the knowns and unknown and how the unknown can be calculated.

   If I know the whole, I can find one part by subtracting the other part from the whole.

- Once pupils are fluent with describing how to work out the function of addition machines, introduce them to a subtraction function machine.

  11 → ? → = 7

- Ask pupils: *What is the same and what is different?*
- Establish that, with this function machine, the number 11 has gone in, and the number 7 has come out. Ask them what they notice about the numbers. What must be happening in this machine if the number that comes out of the function machine is less than the number that

Chapter 3 Numbers up to 20 and their addition and subtraction

Unit 3.9 Practice Book 1A, pages 100–102

went in? Use the bar model to model the calculation and demonstrate that the whole (minuend) is known, as is one of the parts, but you have to work out the other part.

- In pairs, pupils rehearse their explanation of what the function of this machine is and how they know. Explain that it is important to check that they have identified the function correctly by testing it with further numbers. Repeat with other examples.

- Pupils should complete Question 4 in the Practice Book.

Look out for … pupils who are using an inefficient strategy (such as using trial and improvement approaches) because they are unable to use their knowledge of the relationships between parts and wholes to answer the problem.

Same-day intervention

- Use coloured number tiles and a piece of cardboard and set out a calculation such as 5 + 3 = 8. (Small numbers are used to help make it easier to understand the relationships between the numbers rather than only calculating the answers.)

- Cover over the 3 so that the calculation is presented as an empty box missing number problem (5 + ☐ = 8). Ask pupils to read the question: 5 added to something is equal to 8. Ask pupils: *What do you add to 5 to equal 8?*

- Ask pupils to use coloured number rods to model the calculation. Ask them to identify which of the rods were the known values in the calculation and which one shows us what is in the empty box.

- Remind pupils that the model they have made is the same as the model they made when the rods were used for subtraction and ask them what strategy they would use to help them find the missing number. They may count up on their fingers 6, 7, 8 and identify that the answer is 3.

- Ask pupils what the whole is (8) and what the difference is between 8 and 5 (the known addend). Explain that the unknown value is the difference between the whole and the part, which is what we find when we subtract. Ask pupils to use subtraction to find the difference.

- Repeat with other simple calculations until pupils are more fluent about explaining what they know and what they need to find out, and are using a reliable strategy based on reasoning rather than trial and improvement.

Same-day enrichment

- In pairs, pupils draw function machines for their partners to work out the function. These might include function machines that start with the total or difference.

Challenge and extension question

Question 5

5 Choose six of these numbers and then use them to fill in the boxes so that the equations are correct.

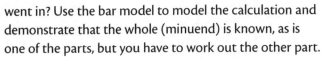

☐ + ☐ = ☐ + ☐ = ☐ + ☐

This task takes the concept of equivalence modelled previously, using the chairs activity, and extends it to include multiple calculations rather than focusing on balancing the addends and their total. If necessary, repeat and extend the chairs task so that pupils retain and develop their understanding of equivalence and knowledge that it applies to multiple calculations.

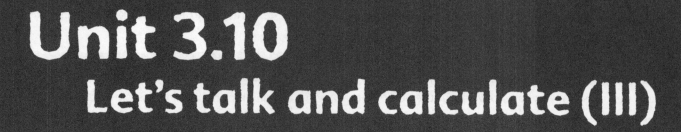

Chapter 3 Numbers up to 20 and their addition and subtraction

Unit 3.10
Let's talk and calculate (III)

Conceptual context

In this unit, pupils will have opportunities to apply addition and subtraction strategies and number facts they have learned to word problems. This is an important part of learning about mathematics as it forges explicit links between classroom mathematics and their experiences of addition and subtraction situations. When solving problems, pupils have to go through several phases: comprehending the text, translating the context into a mathematical model, executing and evaluating the answer. In order to become fluent in problem solving, pupils must be proficient in all of these steps.

Research has found that the wording of mathematical problems can have a significant effect on pupils' ability to solve the problems correctly, even when the actual mathematics demanded is within pupils' capability. In order to develop their ability to interpret and model questions appropriately, in this unit pupils are introduced to some of the different types of problems (change, compare and combine) initially in pictorial form so that they can more easily visualise the problem and model the solution, and then in words.

Learning pupils will have achieved at the end of the unit

- Pupils will have been able to identify the most appropriate strategy for solving different types of word problem (Q1, Q2, Q3)
- Pupils will have developed their proficiency in modelling, recording and explaining their solutions to word problems (Q1, Q2, Q3)
- Pupils will have developed their understanding of part–part–whole relationships in addition and subtraction (Q1, Q2, Q3)

> ⓘ Much research has focused on the structure of word problems and has categorised them into three types. Although the labels vary, researchers are generally agreed that the types are 'combine', 'change' and 'compare' (Riley and Greeno, 1983; 1988). The level of difficulty in each type of question varies, depending on the position of the known and unknown elements.
>
> Combine problems reflect the aggregation structures of addition wherein two or more parts are combined to make a whole or total. Complexities arise when the whole is known and one or other part is unknown and pupils have to apply their knowledge of the inverse relationships between addition and subtraction to reach a solution.
>
> Change problems can reflect the reduction structure of subtraction wherein the subtrahend is removed from the minuend to leave the difference or the augmentation structure of addition wherein some more are added to the original quantity. Complexities arise when the difference is known and the minuend or subtrahend is unknown or when the total is known but not one of the addends. This can then require pupils to apply their knowledge of the inverse relationships between addition and subtraction to reach a solution.
>
> Compare problems reflect difference situations wherein pupils have to calculate how much bigger or smaller one value is than the other. Complexities arise when the difference is known and the minuend or subtrahend is unknown and pupils have to apply their knowledge of the inverse relationships between addition and subtraction to reach a solution. Pupils also have to decide whether a counting up or counting back strategy is most appropriate.
>
> Riley, M. S., & Greeno, J. G. (1988). Developmental analysis of understanding language about quantities and of solving problems. *Cognition and Instruction*, 5, 49–101.

Resources

bar models; counters; counting resources (e.g. buttons); card strips to represent bars ; mini whiteboards; 0–20 number lines

Vocabulary

part, whole, addend, minuend, subtrahend, known, unknown, combine, change, compare

Chapter 3 Numbers up to 20 and their addition and subtraction   Unit 3.10 Practice Book 1A, pages 103–105

Question 1

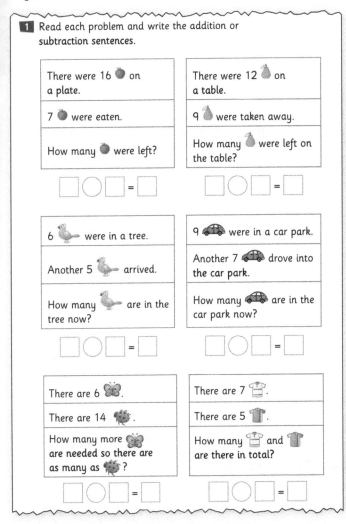

What learning will pupils have achieved at the conclusion of Question 1?

- Pupils will have been able to identify the most appropriate strategy for solving different types of word problem.
- Pupils will have developed strategies suitable for solving different types of word problem structures.
- Pupils will have developed their proficiency in modelling, recording and explaining their solutions to word problems.
- Pupils will have developed their understanding of part–part–whole relationships in addition and subtraction.

Activities for whole-class instruction

- The task in Question 1 gives pupils practice with the three types of addition and subtraction word problems (combine, change and compare). To become familiar with the different types of problem, pupils should have experience of 'acting them out' to help recognise the relationship between the wording of the problem and its solution.
- For each one, pupils will use small objects and bar models to represent the mathematics within the problem to help them understand what they need to calculate.
- 'Combine' questions. Lay out 12 buttons on a table for pupils to see. Say and demonstrate: *There are 12 buttons on the table and then I am given 3 more. How many buttons will I have altogether?* Explain to pupils that this question asks you to combine two parts – the number of buttons that we had at the beginning and the three more buttons given – to find out how many there are altogether.
- Ask pupils, in pairs, to write the addition sentence that matches this question. Do pupils understand that the parts are known and they have to calculate the total?
- Use bar models to model and record the question. Start by using three pieces of card or paper strips to represent the two parts and the whole. Remind pupils that we know the two parts or addends and have to find the whole (sum). Ask pupils where on the bar model they should write 12, 3 and ? to represent what they know and don't know and match the number sentence to the bar model.

$12 + 3 = ?$

- Adapt the question to help pupils to recognise that the structure of the problem remains the same even if the values change. Ask pupils: *What if there were 15 buttons and then 3 more? What changes and what stays the same?* Pupils should draw a new bar model and write a new number sentence on their mini whiteboards. Explain that this is still a combine question as the two parts are combined to find the whole.
- Provide pairs of pupils with some small objects and some combine-type questions where two or more quantities are added together. With a partner, pupils should act out the question to help identify the question type. Pupil 1 uses the objects to model a combine word problem. Pupil 2 draws the matching bar model and writes the appropriate number sentence to match the problem.
- 'Change' questions. Lay out 15 buttons on the table for pupils to see. Say and demonstrate: *There are 15 buttons on the table and 3 were taken away. How many buttons do I have now?*
- Explain to pupils that the situation has now changed: *We had 15 buttons, now that has changed because one part has been taken away and there are 12 buttons remaining.*

207

Chapter 3 Numbers up to 20 and their addition and subtraction   Unit 3.10 Practice Book 1A, pages 103–105

Pupils should write the subtraction sentence that matches this question. Can they explain that the minuend and subtrahend are known and they have to calculate the difference?

- Use bar models to model and record the question. Start by using three pieces of card or paper strips to represent the whole (minuend) and the subtrahend (one part). It will help to place the buttons on the bars too. Physically enact the 'taking away' of the 3 part so that pupils can 'see' that they need to subtract 3. Ask pupils where on the bar model they should write 15, 3 and ? to represent what they know and don't know and match the number sentence to the bar model.

15	
?	3

  $15 - 3 = ?$

- Adapt the question to help pupils recognise that the structure of the problem remains the same even if the values change. Ask: *What if there were ten buttons and then three were removed? What changes and what stays the same?* Pupils should draw a new bar model and write a new number sentence. Check that they understand this is still a change question. Can they explain why?

- Provide pairs of pupils with some small objects and some change questions. With a partner, pupils should act out the question to help identify the question type. Then Pupil 1 uses the objects to model a change word problem. Pupil 2 draws the matching bar model and writes the appropriate number sentence to match the problem.

- 'Compare' questions. Lay out a row of 15 buttons on the table for pupils to see and next to it a row of three buttons. Say and demonstrate: *There are 15 buttons in the first group and three buttons in the second group. How many more buttons are in the first group?* and *How many more buttons do I need to add to the second group to have the same number of buttons as there are in first one?* Explain to pupils that this question asks you to compare the quantities that you have and to work out the difference between them.

- Ask pupils whether they would count on or count back to find the answer to this question and explain that both approaches can be used; which is more efficient will depend on the numbers in the calculation.

- Use bar models to model and record the question. Start by using three pieces of card or paper strips to represent the whole (minuend) and the subtrahend (one part). Remind pupils to find the difference between the minuend and the subtrahend, we can subtract the subtrahend from the minuend – subtract one part from the whole. Ask pupils where on the bar model they should write 15, 3 and ? to represent what they know and don't know and match the number sentence to the bar model.

15	
3	?

  $15 - 3 = ?$

- Adapt the question to help pupils to recognise that the structure of the problem remains the same even if the values change. Ask pupils: *What if there were 15 buttons in the first group and then 12 in the second group? What changes and what stays the same?* Pupils should draw a new bar model and write a new number sentence. Explain that this is still a compare question. Can pupils explain why?

- Provide pairs of pupils with some small objects and some compare questions. With a partner, pupils should act out the question to help identify the question type. Then Pupil 1 uses the objects to model a compare word problem. Pupil 2 draws the matching bar model and writes the appropriate number sentence to match the problem.

- Pupils should complete Question 1 in the Practice Book.

Same-day intervention (for Question 1, Question 2, Question 3)

- Provide pupils with a set of combine, change and compare questions that involve small sets of objects. For each question, they should read it out loud and select objects to represent the question.

- With a partner, they act out the question to help identify the question type and then sort the questions into three groups: combine, change or compare.

- If time allows, once each type of question has been identified, pupils work with a partner to write a new word problem to match each of the types of question.

Same-day enrichment

- Each problem type can be made more complicated by changing what is known and unknown in each question.

- Give pupils a set of calculations and, for each calculation, ask them to develop a word problem that matches it.

  $14 + ? = 18$ (combine) $? + 12 = 15$ (combine)
  $15 - ? = 9$ (change) $? - 8 = 7$ (change)
  $18 - ? = 16$ (compare) $? - 13 = 4$ (compare)

Chapter 3 Numbers up to 20 and their addition and subtraction

Unit 3.10 Practice Book 1A, pages 103–105

Question 2

> 2 Read each problem and then write the number sentence.
>
> (a) There were 6 lambs in the field. Another 7 lambs joined them. How many lambs are there altogether?
>
> Number sentence: _____
>
> (b) Theo has 6 yellow pencils. He has as many blue pencils as yellow pencils. How many pencils does he have altogether?
>
> Number sentence: _____
>
> (c) Orla's sister gave her 12 cherries. She ate 3 of them. How many cherries were left?
>
> Number sentence: _____
>
> (d) 14 children were playing in the park. 8 of them went home. How many children stayed in the park?
>
> Number sentence: _____

What learning will pupils have achieved at the conclusion of Question 2?

- Pupils will be able to identify the most appropriate strategy for solving different types of word problem.
- Pupils will have developed their proficiency in modelling, recording and explaining their solutions to word problems.

Activities for whole-class instruction

- Pupils have developed their understanding of the different structures of word problems through using physical resources and recording the problems with bar models and number sentences.
- Pupils can now move to problems presented in words rather than pictures. They will continue to solve the same range of types of word problem, but now the way in which they are presented has changed.
- Pupils will also revisit and develop their use of number lines as a tool to solve word problems.
- Revisit the whole-class instruction task for Question 1 to develop pupils' fluency with identifying question types, this time changing the quantities to 9, 6 and 15. As well as drawing the bar model and writing the number sentence, pupils should draw a number line to demonstrate the problem type and aid its solution. The variation in representation type will help pupils to make connections between aspects of mathematics as well as provide them with a repertoire of efficient problem-solving strategies.
- For example, when modelling combine questions: lay out nine buttons on a table for pupils to see. Say and demonstrate: *There are nine buttons on the table and then I am given six more. How many buttons will I have altogether?* Pupils use bar models to model and record the question, and then on mini whiteboards, draw the corresponding number line.

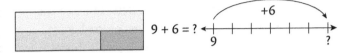

- When modelling change questions, lay out 15 buttons on the table for pupils to see. Say and demonstrate: *There are 15 buttons on the table and nine were taken away. How many buttons do I have now?* Pupils use bar models to model and record the question.

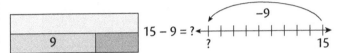

- For each question type, ask pupils which representation they find helps them to identify what they have to find out.
- Pupils complete Question 2 in the Practice Book.

Same-day intervention

- Give pupils some simple examples of each question type.

Helen has 13 shells, Joe has 6 shells. How many shells do they have altogether?
Helen had 15 pens, she threw 5 away. How many pens did she have left?
Helen had 12 grapes, Joe has 4 grapes. How many more grapes does Helen have more than Joe?
Helen has 14 toy animals, Joe has 3 fewer toy animals than Helen. How many toy animals does Joe have?
Helen had 8 biscuits, she gave some to Joe. Helen then had 6 biscuits, how many did she give to Joe?
Helen has 9 apples, Joe has 6 apples. How many apples do they have altogether?
Helen has 14 pencils, Joe has 5 pencils. How many pencils do they have altogether?
Helen has 4 rubbers, Joe has 5 more rubbers than Helen. How many rubbers does Joe have?

Chapter 3 Numbers up to 20 and their addition and subtraction   Unit 3.10 Practice Book 1A, pages 103–105

- In pairs, pupils read each question aloud, acting it out as necessary, and then place each question under the heading of 'combine', 'change' or 'compare'. Once the questions have been sorted, discuss with pupils what helped them to identify the problem type; was it key words (altogether, take away, more/less) or was it the action of combining, taking away or comparing that helped? For each group of questions, pupils should then select the bar model that matches the question.

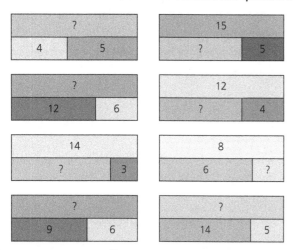

Same-day enrichment

- Provide pupils with opportunities to solve more complex versions of each problem type using bar models or number lines to solve them.

| On Monday, Joe had 10 marbles. On Tuesday, Tom gave him 4 and, on Wednesday, Ali gave him 3 more. How many marbles does Tom have now? (combine) |
| On Thursday, Joe has 7 marbles. On Friday, Tom gave him 5 marbles. On Saturday, Tom gave Ali 4 marbles. How many does Tom have now? (combine/change) |
| Joe had 18 marbles. He gave Tom some marbles. Now Joe has 12 marbles. How many marbles did he give Tom? (change) |
| Joe had some marbles. Then Tom gave him 7 marbles. Now Joe has 18 marbles. How many marbles did Joe have at the beginning? (change) |
| Joe had some marbles. He gave 5 marbles to Tom. Now Joe has 13 marbles. How many marbles did Joe have in the beginning? (change) |

Question 3

> **3** Read each problem and then write the number sentence.
> (a) Jake needs to make 12 cakes. He has made 8. How many more does he need to make?
> Number sentence: _____
> (b) There are 9 grapes left after Tom has eaten 6 grapes. How many grapes were there at first?
> Number sentence: _____

What learning will pupils have achieved at the conclusion of Question 3?

- Pupils will be able to identify the most appropriate strategy for solving different types of word problem.
- Pupils will have developed strategies suitable for solving different types of word problem structures.
- Pupils will have developed their proficiency in modelling, recording and explaining their solutions to word problems.

Activities for whole-class instruction

- Question 2 focused on the combine (aggregation structure of addition) and change (augmentation structure of addition and reduction structure of subtraction) word problem types; Question 3 focuses on the compare type wherein pupils have to find the difference between two quantities.
- In order to help pupils recognise the structure of these types of problem and to help them recognise how they are similar to others, once again pupils will draw bar models, number lines and write number sentences to match each problem. Pupils can then discuss which representation helps them to understand and to solve each problem.
- Focusing on compare questions: lay out a row of 15 buttons on the table for pupils to see and next to it a row of nine buttons. Say and demonstrate: *There are 15 buttons in the first group and nine buttons in the second group. How many more buttons are in the first group?* and *How many more buttons do I need to add to the second group to have the same number of buttons as there are in first one?* Explain to pupils that this question asks you to compare the quantities that you have and work out the difference between them. Ask: *What are the different ways we could do that? What is the best way?*

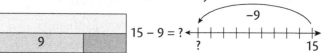

Chapter 3 Numbers up to 20 and their addition and subtraction

Unit 3.10 Practice Book 1A, pages 103–105

- Pupils complete Question 3 in the Practice Book.

Same-day enrichment

- Provide pupils with opportunities to solve more complex versions of each problem type using bar models or number lines to solve them.

Joe has 15 marbles. Tom has 9 marbles. How many fewer marbles does Tom than Joe? (compare)
Joe has 13 marbles. Tom has 5 more marbles than Joe. How many marbles does Tom have? (compare)
Joes has 12 marbles. He has 5 more marbles than Tom. How many marbles does Tom have? (compare)
Joe has 11 marbles. He has 5 fewer than Tom. How many marbles does Tom have? (compare)

Challenge and extension question

Question 4

4 Add a condition to complete the story. Then write the number sentences and find the answers.

(a) There are 8 red pens.

_____.

How many red and green pens are there altogether?

_____

(b) _____.

9 cars drove away. How many cars are left?

_____

Pupils will have developed their understanding of different word problem types and will be able to solve simple problems efficiently. The challenge and extension question deepens this understanding by asking pupils to write a problem and the number sentence to match it.

Chapter 3 Numbers up to 20 and their addition and subtraction

Unit 3.11
Adding on and taking away

Conceptual context

This unit is the penultimate one in the chapter and provides pupils with further opportunities to consolidate and develop their understanding of part–part–whole relationships, which are a crucial part of understanding addition and subtraction calculation strategies. A secure understanding of these relationships helps pupils to identify when to apply a counting up or a counting back strategy to solve a problem. Pupils' understanding is deepened through applying their knowledge to function machines and word problems.

Learning pupils will have achieved at the end of the unit

- Pupils will have consolidated their understanding of the inverse relationship between addition and subtraction (Q1, Q2, Q3)
- Pupils will have consolidated their understanding of part–part–whole relationships in addition and subtraction (Q1, Q2)
- Pupils will have used their knowledge of addition facts to solve subtraction problems (Q1, Q2, Q3)
- Pupils will have reinforced their use of appropriate language to describe part–part–whole relationships (Q1, Q2)
- Pupils will have applied their understanding of inverse relationships to addition and subtraction function machines (Q3)
- Pupils will have greater fluency with their range of known facts through quickly deriving facts from given ones (Q4)

Resources

double-sided counters; ten-frames; interlocking cubes; 0–9 digit cards; envelopes labelled 3, 7, 8, 13, 14; mini whiteboards

Vocabulary

part–part–whole, addend, total, minuend, subtrahend, difference, inverse, doing and undoing

Chapter 3 Numbers up to 20 and their addition and subtraction Unit 3.11 Practice Book 1A, pages 106–109

Questions 1 and 2

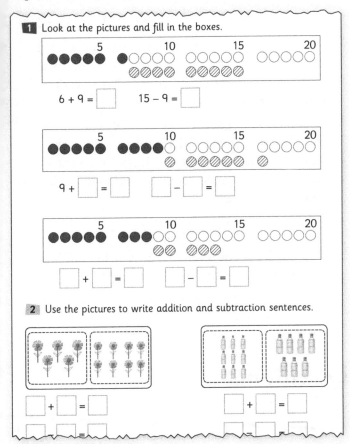

and up to 20 double-sided counters. Ask an addition question, such as 9 + 7 = ? Pupils use the ten-frames and counters to model it and write the addition sentence below.

9 + 7 = 16

- Ask questions to develop pupils' fluency with the language of part–part–whole and addend.

- Pupils should then turn over one part of the calculation (the second addend) to show the subtraction statement: 16 – 7 = 9.

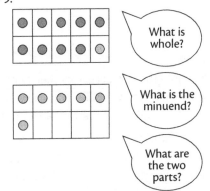

- Ask questions to develop pupils' fluency with the language of part–part–whole and minuend, subtrahend and difference.

- Remind pupils that addition and subtraction are the inverse of each other, which means that each one has the opposite effect of the other, so that one operation reverses or undoes the other. Adding 9 and 7 combines two parts to make the whole or total of 16. Subtraction is the inverse of addition so it separates the whole into two parts.

- Repeat these steps with other calculations such as 7 + 8 = 15 and 9 + 8 = 17 to help develop pupils' fluency with modelling and describing addition and subtraction statements.

- Pupils should complete Questions 1 and 2 in the Practice Book.

What learning will pupils have achieved at the conclusion of Questions 1 and 2?

- Pupils will have consolidated their understanding of the inverse relationship between addition and subtraction.
- Pupils will have consolidated their understanding of part–part–whole relationships in addition and subtraction.
- Pupils will have used their knowledge of addition facts to solve subtraction problems.
- Pupils will have reinforced their use of appropriate language to describe part–part–whole relationships.

Activities for whole-class instruction

- Pupils should already be familiar with using bar models, number lines and ten-frames to represent relationships and calculations, so this whole-class task serves as a recap of previous units. It ensures that all pupils are fluent in both describing and representing part–part–whole relationships.

- Identifying parts and wholes in addition and subtraction. Pupils work with a partner. Each pair has two ten-frames

Chapter 3 Numbers up to 20 and their addition and subtraction   Unit 3.11   Practice Book 1A, pages 106–109

All say... ... If I know that 7 + 8 = 15, then 15 – 7 must equal 8 because subtraction is the inverse of addition.

Look out for ... pupils who are not using their knowledge of inverses to help solve problems (for Question 1, Question 2).

Same-day intervention (Question 1, Question 2)

- It is important that pupils have sufficient first-hand experience of the inverse relationship between addition and subtraction in order to develop greater fluency with their range of known facts through quickly deriving facts from given ones.
- Starting with small numbers so that pupils can focus on the relationships between part and whole, give pupils a set of three red interlocking cubes and a set of five yellow interlocking cubes. Using a number line, show that 3 + 5 = 8.

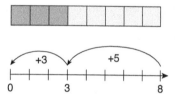

- Ask pupils to join the two parts together to make the whole. Ensure that they recognise that there are two parts (addends) 3 and 5 and these make a total of 8.
- Ask: *How many are in the whole? What is the value of the red part? ... the yellow part?* From the total of 8, ask pupils to take away the five yellow cubes. Explain that what they have done is taken one part from the 8, so what is left (the difference) is the other part, 8 – 5 = 3.

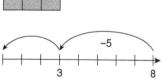

- Repeat with other numbers less than 10, each time ensuring that pupils are joining together the two parts to make the whole and then separating them again into parts to find the difference. Once pupils are confidently working with numbers less than 10, move on to numbers greater than 10.

Same-day enrichment (Question 1, Question 2)

- What could the numbers be? This problem utilises pupils' knowledge of addition and subtraction facts, as well as encouraging them to work systematically to identify pairs of numbers that would give a particular total.
- In pairs pupils are given a set of ten cards, each showing one of the digits from 0 to 9. Explain to pupils that they have to put two digit cards in each envelope and they must add them together to make the total shown on the front of the envelope.

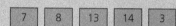

- What numbers could be inside the '8' envelope? What numbers are inside your '8' envelope? Why?
- Ask pupils to explain their strategies for getting started and how they kept track of the numbers used and worked systematically.

Question 3

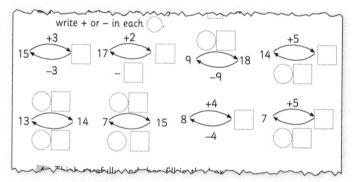

write + or – in each ◯.

What learning will pupils have achieved at the conclusion of Question 3?

- Pupils will have consolidated their understanding of the inverse relationship between addition and subtraction.
- Pupils will have used their knowledge of addition facts to solve subtraction problems.
- Pupils will have applied their understanding of inverse relationships when using addition and subtraction function machines.

Activities for whole-class instruction

- In the previous task, pupils explored part–part–whole relationships using physical and visual representations of them. Pupils will develop their understanding of

Chapter 3 Numbers up to 20 and their addition and subtraction   Unit 3.11   Practice Book 1A, pages 106–109

these relationships and how addition is the reverse of subtraction through applying what they have learned to working with two-direction function machines.

- Pupils will have seen function machines in Chapter 2 and will be familiar with them. Show pupils the function machine showing the top function only (the addition) to begin with. Ask: *What is the machine doing?* (+3)

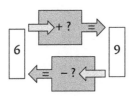

- Reveal the second (inverse) function and explain that this function machine is able to undo the calculations too. Ask: *What will the machine need to do to undo the operation?* (–3) and remind pupils of the:

 Because subtraction is the inverse of addition, then subtracting 3 reverses the +3.

- Provide pupils with other calculations to practise using the two-direction function machines:

  7 ? 15

  9 ? 18

  11 ? 16

- Ask pupils what they see and what they notice about the numbers and operations of each function. Do they notice that one part and the whole are given and they need to find the value of the other part?

- Remind pupils that the two functions reflect that addition and subtraction are inverses, so one calculation will be the reverse of the other. Once pupils are able to explain the basic two-direction function machine and work out the second function, increase the challenge by changing the information provided about the function machine.

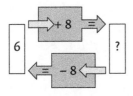

- Ask pupils what is the same and what is different about these function machines and the first ones they looked at. Ensure that pupils notice that in these function machines, the two parts are known but not the total or minuend.

- Ask: *What do you know about part–part–whole relationships that will help you work out the missing parts of the function machine?*

- Pupils complete Question 3 in the Practice Book.

 Because subtraction is the inverse of addition, then subtracting a number reverses adding that number.

Same-day intervention

- Some pupils may need to work with concrete resources for longer in order to recognise the relationships between adding and subtracting from a given number.

- Using counters, make a row of 10. Ask pupils to split the row into two parts and say how many are in each part. Record that the two parts are equal to 10, and write the matching number sentence, for example 8 + 2 = 10.

- From the group of 10, remove the two yellow counters and ask how many are left. Explain that you know that 8 are left because you know that 8 added to 2 is equal to 10. Recombine the two parts and remove the eight counters. Ask pupils how many counters you have removed and how they know. Explain that 2 + 8 = 10, so if 2 remains 8 must have been removed.

- In pairs, pupils make other pairs of numbers to make 10 and identify and record the two parts that have been added to make the whole. They should then remove one part and explain to their partner how they know the value of the other part.

Same-day enrichment

- Pupils can add to the complexity of the function machines by having two steps to each function. They have to not only find the missing information but also convert a two-step function in one direction to a single-step reverse function. Once they have solved the ones provided, pupils can develop their own.

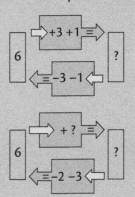

Chapter 3 Numbers up to 20 and their addition and subtraction   Unit 3.11   Practice Book 1A, pages 106–109

Question 4

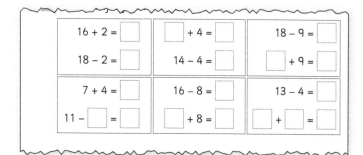

What learning will pupils have achieved at the conclusion of Question 4?

- Pupils will have greater fluency with their range of known facts through quickly deriving facts from given ones.

Activities for whole-class instruction

- Most pupils should have a secure understanding of the inverse relationships between addition and subtraction. This task focuses on developing fluency in recall of known facts and using known facts to quickly derive new ones.

- If I know one fact, what else do I know? Pupils should each have a mini whiteboard. The aim is to try and recall facts instantly or derive them quickly. Give the class an addition or subtraction question and, as quickly as they can, pupils write down the corresponding addition and subtraction facts. After 30 seconds (adapt as necessary), say *Show me* so everyone can see the related facts.

- Ask one pupil to explain what the parts and wholes are in their number sentences:

7 + 5 = 12
6 + 8 = 14
14 − 8 = 6
17 − 9 = 8
? + 7 = 15
? + 6 = 12
4 + ? = 11
6 + ? = 13
? − 6 = 10
? − 7 = 12
14 − ? = 5
13 − ? = 8

- Pupils complete Question 3 in the Practice Book.

Same-day enrichment

- Knowledge of inverses is often useful when checking calculations for accuracy. Provide pupils with a set of calculations, some of which have been answered incorrectly. For each calculation, pupils work out the inverse calculation to use to check the one given and identify the incorrect answers.

Original calculation	Inverse calculation	Correct or incorrect?
13 + 6 = 20		
7 + 8 = 15		
9 + 12 = 19		
15 − 9 = 7		
18 − 6 = 8		
14 − 8 = 6		

Chapter 3 Numbers up to 20 and their addition and subtraction   Unit 3.11   Practice Book 1A, pages 106–109

Challenge and extension question
Question 5

5 Read each number story and then write the question. Then write the number sentence.

(a) There were 18 children on the train. 4 of them got off at a station. _____?

Number sentence: _____

☐ ◯ ☐ = ☐

(b) Min bought 15 books, and then she bought another 4 books. _____?

Number sentence: _____

☐ ◯ ☐ = ☐

(c) There were 14 cars in the car park at first. 8 of them then drove away. _____?

Number sentence: _____

☐ ◯ ☐ = ☐

(d) Amber had 13 strawberries. She gave 8 to Jacob. _____?

Number sentence: _____

☐ ◯ ☐ = ☐

Pupils have revised and developed their understanding of part–part–whole relationships and inverses through using concrete resources, pictures and symbols. This knowledge is now applied to word problems wherein pupils have to identify the calculation from the contexts provided. They are then further challenged to change the word problem to reflect the inverse number sentence too.

Chapter 3 Numbers up to 20 and their addition and subtraction

Unit 3.12
Number walls

Conceptual context

Pupils have had extensive experience of working with numbers up to 20, developing fluency and instant recall of facts as well as using a known fact and their understanding of inverses and part–part–whole relationships to quickly derive new facts. The inclusion of number pyramids provides an effective end-of-unit application of this knowledge as, for each missing number, pupils have to identify the information they have, and how they will use it to find the unknown part or whole. They also have to decide whether to use addition or subtraction (or known fact knowledge) to work out the missing values.

Learning pupils will have achieved at the end of the unit

- Pupils will have developed fluency with number bonds to 20 (Q1, Q2)
- Pupils will have extended their application of their understanding of inverses to identify appropriate calculation strategies (Q1, Q2)
- Pupils will have developed strategies for solving problems efficiently (Q1, Q2)
- Pupils will have reinforced their use of mathematical language by explaining their solutions to problems (Q1, Q2)

Resources

1–20 number cards; mini whiteboards

Vocabulary

part–part–whole, addend, total, minuend, subtrahend, difference

Chapter 3 Numbers up to 20 and their addition and subtraction   Unit 3.12 Practice Book 1A, pages 110–111

Questions 1, 2 and 3

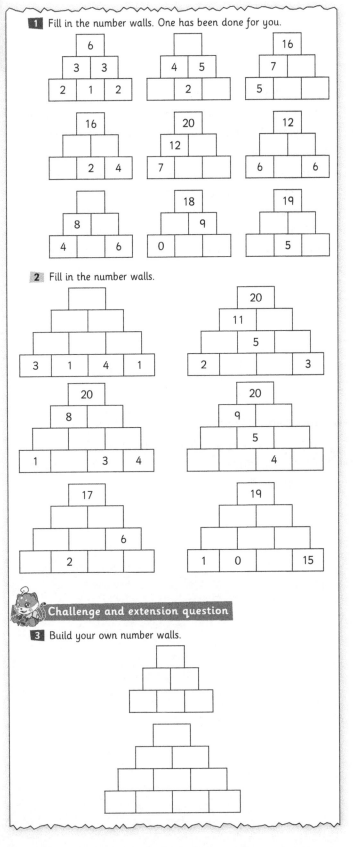

What learning will pupils have achieved at the conclusion of Questions 1, 2 and 3

- Pupils will have developed fluency with number bonds to 20.
- Pupils will have extended their application of their understanding of inverses to identify appropriate calculation strategies.
- Pupils will have developed strategies for solving problems efficiently.
- Pupils will have reinforced their use of mathematical language by explaining their solutions to problems.

Activities for whole-class instruction

- What parts could make this whole? Say a number between 10 and 20 and say: *This is the whole, what could the parts be?* In pairs, pupils write on whiteboards as many possible parts to make this whole as possible. Pupils should also be encouraged to be systematic and to identify any pairs of numbers they have written twice.
- Repeat this task so that pupils are recalling instantly (or deriving quickly) appropriate parts and are working systematically and efficiently to find all of the possibilities.
- What is the minuend? Give pairs of pupils two numbers, at least one of which must be less than 10. Ask pupils to work out the minuend if these two numbers are the parts.

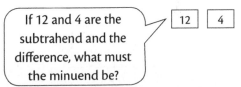

- Place a digit card on the board for everyone to see. Place two more cards to show the arrangement below. Ask pupils to watch closely and ask them what they notice about the three cards used so far. Explain that the 8 is the sum of the two numbers below it.

- Add a row below the 3 and 5 and ask pupils what they notice. Point out that the row below 8 are the addends that total 8 (3 + 5) and the row below 3 and 5 are the addends that total 3 (1 + 2) and 5 (2 + 3). Explain that the 2 is used in both calculations. Explain that the pyramid can be checked through using knowledge of both addition and subtraction bonds: 3 + 5 = 8 and 8 − 5 = 3:

219

Chapter 3 Numbers up to 20 and their addition and subtraction   Unit 3.12 Practice Book 1A, pages 110–111

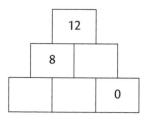

```
      8
    3   5
  1   2   3
```

- Start a new pyramid. Position the number at the top, and ask pupils what the two numbers below could be (utilising the 'what could the parts be?' task). Ensure all possible combinations have been used, then position the addends below the 7. Ask pupils how many numbers will be below the 3 and 4 and ask them to find the three missing values.

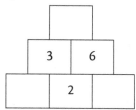

- Ask pupils if any of them found another set of numbers that would work. Ask them to explain to the class how they worked out the missing values and how they checked.

- Present pupils with a pyramid where what is known and unknown has changed. Ask pupils to say what they know already and to explain what they have to find out. Each pair then try to complete the pyramid and have to convince another pair that their solution is correct.

```
    3   6
      2
```

- Ask pupils: *What happens if the arrangement of known and unknown numbers change?* Present pupils with a pyramid with different empty squares.

- Ask pupils to say what they know already and to explain what they have to find out. Each pair then try to complete the pyramid and have to convince another pair that their solution is correct.

- Repeat this process for pyramids with a fourth row. Each time, each pair of pupils must identify what they know and what they don't know and of they will use addition or subtraction facts to find the missing numbers. Encourage pupils to be systematic when trying pairs of numbers.

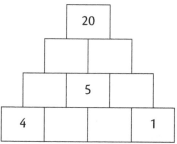

- Pupils complete Questions 1, 2 and 3 in the Practice Book.

Look out for ... pupils who find it difficult to think of a second set of numbers to make a total if the first pair doesn't work in the pyramid.

Chapter 3 Numbers up to 20 and their addition and subtraction   Unit 3.12 Practice Book 1A, pages 110–111

Same-day intervention

- This unit focuses on developing pupils' fluency and instant recall with addition and subtraction facts to 20. Pupils will find it very difficult to memorise facts unless they have conceptual understanding of addition, an understanding that the order of the addends does not change the sum (the commutative principle) and they can solve addition questions correctly under un-timed conditions. In order to be able to memorise subtraction facts, pupils need to have conceptual understanding of subtraction and the ability to answer subtraction calculations correctly under un-timed conditions.
- It should be noted that pupils can be at different levels for addition and subtraction, with pupils usually being able to memorise addition facts before they can memorise subtraction ones.
- In order to help pupils to make connections between addition and subtraction facts and representations of the relationships between parts and whole, give pupils calculations presented using a ten-frame. For each calculation, pupils must say the equivalent addition calculation and both corresponding subtraction ones. They use the counters to identify the parts and whole (addends, minuends, subtrahends and differences). The focus is on identifying related facts rather than speed.

For example: 8 + 4 = 12, 4 + 8 = 12, 12 − 4 = 8, 12 − 8 = 4

Chapter 3 test (Practice Book 1A, pages 112–114)

Test question number	Relevant unit	Relevant questions within unit
1	3.3	1, 2, 3, 4
	3.4	2
	3.11	1, 2
2	Not specific to unit	
	3.12	1, 2, 3,
	3.11	1
3	3.11	2
	3.9	1
4	Not specific to unit	

Resource 1.1.2

Can you see … ?

Resource 1.1.7a

On the farm

horses =

sheep =

cows =

ducks =

Draw 8 birds flying in the sky.

Resource 1.1.7b

Is this 10?

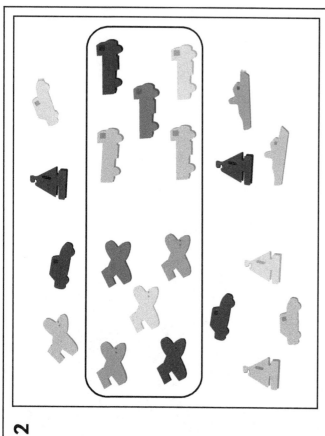

2

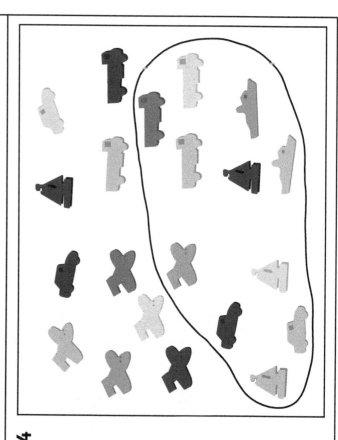

4

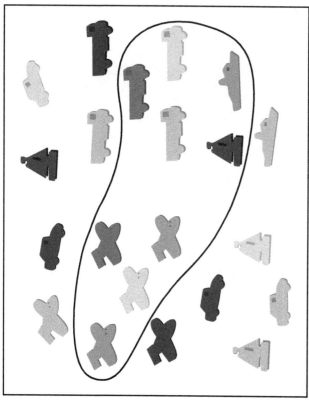

1

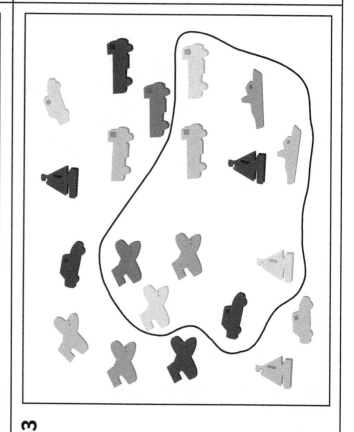

3

Blank spinner

How to use this spinner

Divide the spinner into equal sections and number them. Hold the paper clip in the centre of the spinner using the pencil and gently flick the paper clip with your finger to make it spin.

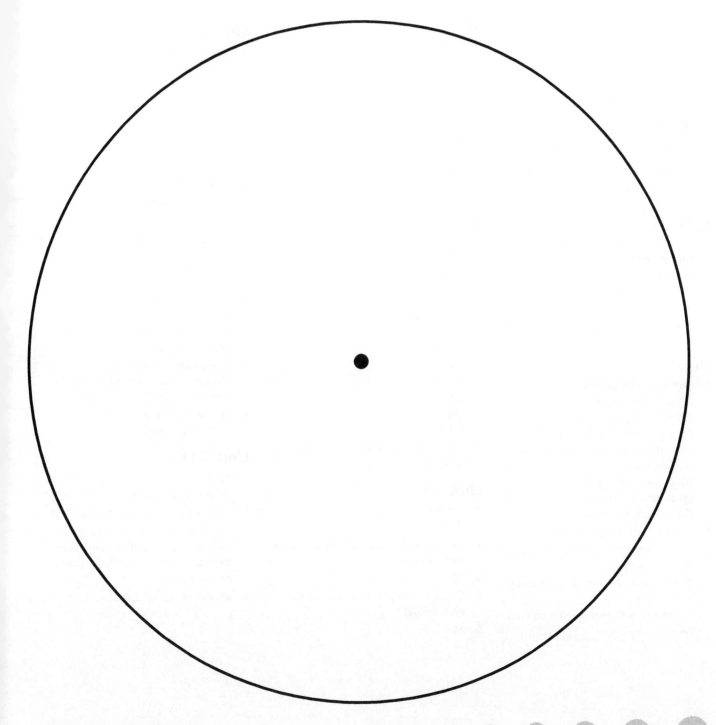

Answers

Chapter 1 Numbers up to 10

Unit 1.1
1. Lines should join:
   bird and eggs
   shoes and feet
   flower and vase
   pencil and notepad
   squirrel and tree
2. Lines should join:
   butterflies and triangles
   apples and circles
   books and hearts
   cars and squares
3. 1, 3, 4, 6, 7, 10, 2, 7, 10, 5
4. Check pupil's shading matches pattern.

Unit 1.2
1. toys: doll, teddy, football, car, helicopter, bricks
   art and craft: pens, crayons, notebook, scissors, paintbrushes, paints
2. Check pupil's answers make sense.
3. There should be a circle around the milk, the vase, the bicycle and the cylinder.
4. In the air: 3, 6, 8
   On land: 1, 4, 7, 10
   In the water: 2, 5, 9
   Alternative groupings may vary. Check that they make sense.

Unit 1.3
1. Sets should be as follows:
   (a) chicks and hens
   (b) girls and boys
   (c) apples and strawberries
   (d) vases of flowers and empty vase
   (e) cats and mice
   (f) presents and cards
   (g) pencils and rulers
   (h) balloons and kites
2. Check pupil's groups make sense.
3. Animals with four legs: dog, monkey, squirrel, horse, cat
   Animals that can climb trees: monkey, squirrel, cat
   Animals in both sets: monkey, squirrel, cat

Unit 1.4
1. 1: watermelon, 2: pineapples, 3: pears, 4: bananas, 5: apples, 6: plums, 7: kiwi fruit, 8: oranges, 9: bunches of grapes, 10: grapes
2. triangles: 8
   circles: 10
   squares: 9

Unit 1.5
1. 3, 1, 2, 5, 4
   two, 2
   three, 3
   one, 1
   five, 5
   four, 4
   8, 9, 10
   seven, 7
   ten, 10
   six, 6
   three, 3
2. triangle, 2
   circle, 3
   square, 1
   rectangle, 3

Unit 1.6
1. Sets should be as follows, in any order:
   1
   2, 3 and 4
   5 and 6
   7 and 8
2. (a) 2, 3, 4
   (b) 2, 2, 6, 3
   (c) 3, 3, 2, 4, 1
3. The first and third towers should be shaded.

Unit 1.7
1. 5, 8, 10
   9, 5, 6
2. Check that 6, 7, 8, 9 and 10 circles have been shaded.
3. 10
4. Check that pupils have circled 10 of each animal.
5. 10

Unit 1.8
1. (a) 3, 2, 1, 0
   (b) 6, 4, 0
2. 5, 9, 8, 0, 3
3. Check that pupils have drawn 10, 0, 7, 4 triangles.
4. 0 cm
5. 0, 5, 10

Unit 1.9
1. 5, 3; 2, 4; 5, 4
   1, 4; 5, 5; 6, 1
2. 0, 5
   1, 4
   2, 3
   3, 2
   4, 1
   5, 0
3. Pupils should add the following numbers of eggs:
   7, 4, 4
   2, 7, 4
4. Check pupil's answers add to 8.
5. 8, 4, 4 and 10, 5, 5

Unit 1.10
1. 5, first, fourth
2. (a) 10, tenth, fourth, first, seventh
   (b) sixth, second, 7, 3, 3, 2
3. Check that five of the hearts are coloured in the first set and the fifth heart is coloured in the second set.
4. (a) 3 white triangles
   (b) 2 white circles

Unit 1.11
1. 10, 5
2. third, fourth, first, second, sixth, fifth
3. fifth, sixth, second, first, fourth, third
4. Check that there is a triangle in the fifth place, a circle in the tenth place, a square in the third place and two hearts in the seventh place.
5. second, third, first, fourth

Answers

Unit 1.12
1. The following sets should be ticked:
   (a) toothbrushes
   (b) shorts
2. The following sets should be ticked:
   (a) strawberries
   (b) triangles
3. 4, 5, 1, 1 (cylinders and cubes)
   3, 7, 4, 4 (stars)
4. Check that pupils have drawn:
   4 circles
   7 triangles
   2 squares
5. Pencil 2 should be coloured.

Unit 1.13
1. > = <
2. 4 = 4, 3 < 5, 4 > 2
3. Pupils should draw more than three circles, less than three triangles, and five rectangles.
4. < > > =
   > < = <
5. Answers will vary. Check that they make sense.
6. The third glass should be circled.

Unit 1.14
1. b should be ticked
2. (a) 2, 3, 4, 5, 6, 7, 9
   (b) 0, 4, 6, 10
3. (a) 3, 2, 6
   (b) 2, 4, 2, 8, 4
4. (a) Check that 7, 0, 3, 2 and 9 are circled.
   (b) 0, 2, 3, 7, 9
   (c) 10, 8, 6, 5, 1
5. Answers will vary. Check that they make sense.

Chapter 1 test
1. Lines should join:
   plant and plant pot
   fork and knife
   scarf and gloves
   monkey and bananas
   cake and present
2. Lines should join:
   strawberries and seven
   bananas and three
   apples and ten circles
   cherries and 4
   oranges and five circles
3. Groupings will vary. Check that they make sense.
4. third, fifth, sixth, seventh, seven
5. (a) The first four apples should be shaded.
   The fourth apple should be shaded.
   (b) The second strawberry should be shaded.
   All except the first strawberry should be shaded.
6. five, 3, second, fourth, 2
7. 3, 2, 1, 4
8. 0, 2, 3, 4, 6, 7, 9, 10
9. (a) 1, 3, 5, 7, 9
   1, 2, 9
   (b) 9, 3, 0
10. squares, 3, circles/triangles, triangles/circles
11. < > = > >
    8
    The remaining answers will vary. Check that they make sense.
12. (a) 2, 4, 5, 8, 9, 10
    (b) 10, 7, 6, 4, 1, 0

Chapter 2 Addition and subtraction within 10

Unit 2.1
1. 0, 4; 1, 4
   1, 3; 2, 3
   2, 2; 3, 2
   3, 1; 4, 1
   4, 0; 5, 0
2. 3, 5
3. 3, 7, 4, 3
   5, 5, 7, 0, 2
4. (a) 5 + 1 = 6
   4 + 2 = 6
   3 + 3 = 6
   2 + 4 = 6
   1 + 5 = 6
   (b) 9 + 1 = 10
   8 + 2 = 10
   7 + 3 = 10 3 + 7 = 10
   6 + 4 = 10 2 + 8 = 10
   5 + 5 = 10 1 + 9 = 10
   4 + 6 = 10

Unit 2.2
1. 8, 9, 5, 10, 10
2. 3 + 4 = 7   3 + 2 = 5   6 + 4 = 10
   4 + 3 = 7   2 + 3 = 5   4 + 6 = 10
   2 + 6 = 8   5 + 4 = 9   1 + 7 = 8
   6 + 2 = 8   4 + 5 = 9   7 + 1 = 8
3. 10, 6, 8, 10, 6
   8, 10, 7, 3, 10
4. Lines should be joining:
   1 + 3 and 2 + 2
   1 + 1 and 0 + 2
   5 + 2 and 1 + 6 and 4 + 3
   3 + 5 and 8 + 0
5. 5 + 0 = 5   2 + 3 = 5
   4 + 1 = 5   1 + 4 = 5
   3 + 2 = 5   0 + 5 = 5
   (These calculations may appear in any order.)

Unit 2.3
1. (a) 0 + 8 = 8   1 + 7 = 8   2 + 6 = 8
       3 + 5 = 8   4 + 4 = 8   5 + 3 = 8
       6 + 2 = 8   7 + 1 = 8   8 + 0 = 8
   (b) Ensure that the coloured dots match the additions. The calculations may appear in a different order.
       9 + 0 = 9   1 + 8 = 9   2 + 7 = 9
       3 + 6 = 9   4 + 5 = 9   5 + 4 = 9
       6 + 3 = 9   7 + 2 = 9   8 + 1 = 9
2. 4 + 3 = 7   5 + 5 = 10
   1 + 7 = 8   6 + 4 = 10
3. 8, 10, 7
   4, 8, 6
   4, 8, 10
   6, 9, 9
4. Answers will vary. Possible answers include:
   2 + 3 = 5
   3 + 4 = 7
   5 + 4 = 9
   4 + 2 = 6

Unit 2.4
1. 6 + 4 = 10
   4 + 4 = 8   6 + 3 = 9
   3 + 4 = 7   3 + 7 = 10
2. 7, 3, 8
   6, 9, 7
   5, 1, 3
   +3, +6, +0
3. 8, 8, 8
   5, 8, 9
   8, 5, 10
   10, 6, 4
   10, 9
   8, 6
4. (a) 6 and 2
   (b) six triangles

Answers

Unit 2.5
1. 6 + 3 = 9   3 + 4 = 7
   5 + 2 = 7   4 + 5 = 9
   3 + 6 = 9   4 + 2 = 6
   4 + 5 = 9 and 5 + 4 = 9 (in any order)
   3 + 6 = 9 and 6 + 3 = 9 (in any order)
2. 7, 6, 8
   8, 9, 10
   10, 7, 10
   7, 6, 1
   3, 5, 9
3. 5, 9, 5, 8
   4, 10, 6, 9
   6, 7, 9, 7
   10, 7, 9, 10
4. Pairs of calculations can be in either order:
   4 + 1 = 5   1 + 4 = 5
   4 + 3 = 7   3 + 4 = 7
   6 + 4 = 10 4 + 6 = 10
   4 + 2 = 6   2 + 4 = 6
   5 + 2 = 7   2 + 5 = 7
   2 + 1 = 3   1 + 2 = 3
   3 + 0 = 3   0 + 3 = 3
   5 + 3 = 8   3 + 5 = 8
   2 + 2 = 4
   4 + 0 = 4   0 + 4 = 4
   2 : 2

Unit 2.6
1. 6 − 2 = 4
   6 − 4 = 2
   7 − 2 = 5
   9 − 3 = 6   10 − 8 = 2
2. 4, 1, 2, 0
   1, 3, 5, 5
   5, 1, 2, 2
   0, 2, 3, 3
3. 6, 6, 2, 8, 0
   6, 0, 1, 1, 2
4. Calculations may appear in any order:
   10 − 5 = 5   7 − 2 = 5
   9 − 4 = 5   6 − 1 = 5
   8 − 3 = 5   5 − 0 = 5

Unit 2.7
1. 2, 4, 2, 3, 8
2. 7 − 4 = 3   10 − 6 = 4
   8 − 5 = 3   10 − 3 = 7
   9 − 7 = 2   7 − 2 = 5
3. 3, 1, 2
   8, 1, 8
   3, 0, 5
   −4, −3, −6
4. 4, 0, 5, 5
   1, 5, 3, 1
   3, 4, 3, 5
   8, 1, 0, 2

5. 2 − 0 = 2   3 − 1 = 2
   4 − 2 = 2   5 − 3 = 2
   6 − 4 = 2   7 − 5 = 2
   8 − 6 = 2   9 − 7 = 2

Unit 2.8
1. 7 − 1 = 6   9 − 5 = 4
   8 − 3 = 5   9 − 3 = 6
   7 − 4 = 3   8 − 5 = 3
   10 − 5 = 5 6 − 6 = 0
2. 4, 5, 3
   3, 4, 6
   7, 3, 10
3. 3, 6, 0, 6
   3, 1, 3, 5
   7, 3, 1, 3
4. 6 + 4 = 10
   6 − 4 = 2
   6 − 4 = 2

Unit 2.9
1. 6 − 2 = 4   10 − 7 = 3
   8 − 5 = 3   7 − 4 = 3
   9 − 4 = 5   7 − 5 = 2
2. 4, 4, 2, 3
   3, 1, 7, 1
   4, 2, 9, 4
   8, 1, 0
3. 3, 9, 0, 4
   10, 6, 7, 7
   9, 3, 1, 1
   3, 10, 4, 2
4. 4, 6
   9, 2

Unit 2.10
1. 5 + 2 = 7   7 − 2 = 5
   6 + 3 = 9   9 − 3 = 6
   10 − 3 = 7 7 + 3 = 10
2. 8, 9, 6, 4
   9, 9, 10, check pupil's answers
3. 5   10   5
   1   6   3
   4   4   2
   9   10   7
   Check pupils' answers.
4. 5, 9, 7, 2
   10, 5, 3, 6
   10, 4, 0, 10
5. Calculations may appear in any order:
   10 − 7 = 3   6 − 3 = 3
   9 − 6 = 3   5 − 2 = 3
   8 − 5 = 3   4 − 1 = 3
   7 − 4 = 3   3 − 0 = 3

Unit 2.11
1. 0, 2, 4, 5, 6, 7, 8, 10
   1, 3, 5, 7, 9

2. 3 + 4 = 7
   0 + 9 = 9
   8 − 6 = 2
3. 8, 6, 6, 9
   4, 10, 10, 2
   9, 10, 10, 7
   5, 0, 2, 4
4. 4, 8, 4
   3, 0, 9
   4, 9, 2
5. Check pupil's number lines.
   2 + 4 = 6
   6 − 5 = 1

Unit 2.12
1. Lines should join:
   1 and 9
   2 and 8
   3 and 7
   4 and 6
   5 and 5
   10 and 0
2. From left numbers should be:
   7, 8, 9, 10, 0, 5, 5, 4
3. 7, 9, 0
   5, 4, 2
4. The following pairs should be joined:
   2 + 8 and 3 + 7   7 − 5 and 6 − 4
   4 + 5 and 2 + 7   9 − 3 and 8 − 2
   2 + 6 and 5 + 3   8 − 4 and 9 − 5
   9 − 2 and 3 + 4   6 − 3 and 7 − 4
5. 5, 10, 1, 9
   7, 8, 3, 10
   2, 4, 4
   Last 2 could be 10, 7 or 9, 6 or 8, 5 or 7, 4 or 6, 3 or 5, 2 or 4, 1 or 3,0
6. >   >   >
   <   =   <
   =   <   =
7. Answers will vary. Check that they all add to 10.

Unit 2.13
1. 2 + 3 + 2 = 7   3 + 4 + 2 = 9
   3 + 5 + 2 = 10 4 + 1 + 3 = 8
2. 1 + 3 + 4 = 8
   0 + 2 + 7 = 9
3. 7, 7, 10
   7, 5, 9
   9, 10, 10
   7, 10, 10
4. (a) 3 + 2 + 5 = 10
   (b) 4 + 2 + 2 = 8
5. 5, 4; 3, 4
   6, 2; 1, 4

Answers

Unit 2.14
1. $10 - 2 - 4 = 4$ $10 - 5 - 3 = 2$
2. $9 - 5 - 2 = 2$
   $7 - 1 - 2 = 4$
3. 1, 0, 2
   2, 4, 1
   0, 3, 0
   4, 2, 2
   2, 0, 0
4. (a) $9 - 5 - 2 = 2$
   (b) $9 - 3 - 4 = 2$
5. 5, 3; 9, 1
   3, 6; 9, 8

Unit 2.15
1. $9 - 4 + 1 = 6$ $9 - 3 + 4 = 10$
   $5 - 1 + 3 = 7$ $9 - 5 + 2 = 6$
2. $2 + 7 - 4 = 5$
   $10 - 6 + 3 = 7$
3. < =
   = >
   > =
4. 9 7
   6 10
   3 2
   1 9
   6 8
   10 7
5. 8, 1; 9, 7
   8, 5; 1, 0

Chapter 2 test
1. 6, 5, 3, 0
   0, 9, 3, 10
   10, 2, 7, 5
2. Lines should join:
   $3 + 4$ and $1 + 6$
   $2 + 2$ and $8 - 4$
   $10 - 1$ and $9 - 0$
3. (a) $+6, 2 + 6 = 8$
   (b) $-6, 7 - 6 = 1$
   (c) $+5, -2, 3 + 5 - 2 = 6$
4. Lines should join:
   $4 + 6 - 0$ and 10
   $2 + 5 + 3$ and 10
   $9 - 2 - 2$ and 5
   $7 - 4 + 7$ and 10
   $5 + 5 - 5$ and 5
   $10 - 1 - 3$ and 6
   $8 - 3 + 0$ and 5
   $9 - 5 + 2$ and 6
5. $4 + 2 = 6$ $9 - 5 = 4$
   $3 + 4 + 3 = 10$ $8 - 4 - 2 = 2$

Chapter 3 Numbers up to 20 and their addition and subtraction

Unit 3.1
1. 13, 12, 16
   13, 19, 11
   19, 15, 20
2. 11, 12, 14, 15, 16, 18, 20
   14 15 16 (any two of these in ascending order)
   20 Remaining two numbers may vary but both should be less than 19, with the first number bigger than the second.
3. 12, 10, 8   10, 5
   15, 13, 11, 7, 5, 3, 1
4. (a) Green line should join 11, 13, 15, 17, 19
   Red line should join 10, 12, 14, 16, 18, 20
   (b) 11, 13, 15, 17, 19
   10, 12, 14, 16, 18, 20

Unit 3.2
1. Line should be joining:
   hearts and 14
   stars and 15
   circles and 11
2. 14   13
   $10 + 4$ $10 + 3$
   16   18
   $10 + 6$ $10 + 8$
   20   17
   $10 + 10$ $10 + 7$

3. $12 - 2 = 10$ $10 + 5 = 15$
   $10 + 10 = 20$ $16 - 6 = 10$
4. 12 15 19
   12 15 19
   8   6   3
   10 10 10
5. 4   7   10
   10 10 10
   2   6

Unit 3.3
1. 3, 4, 6, 8, 9
   12, 14, 15, 17, 19, 20
   1, 4, 5, 6, 10
   12, 13, 14, 16, 17, 20
2. Check that all odd numbered beads are shaded.
   Check that all even numbered beads are shaded.
   Check that 5, 10, 15 and 20 are shaded.
3. 10, 12, 13, 15, 16, 17, 19, 20
   (a) 14, 16
   (b) 19, 10
   (c) 12
   (d) 13, 14, 15, 16, 17
   (e) 16, 17, 18, 19
4. (a) 1, 3, 5, 7, 9
   (b) 20, 19, 18, 17, 16, 15
   (c) 3, 4, 5, 6, 7
   (d) 4, 8, 12, 16, 20

5. (a) 8 should be circled: 3, 5, 7, 9
   (b) 9 should be circled: 2, 4, 6, 8
   (c) 18 should be circled: 17, 15, 13, 11
   (d) 13 should be circled: 6, 9, 12, 15, 18
6. (a) Any one of 13 or 14
   (b) Any one of 17, 18 or 19
   (c) Answers will vary. One possible answer is 3, 5, 7, 9, 11, 13, 15

Unit 3.4
1. $2 + 6 = 8$ $12 + 6 = 18$
   $8 - 6 = 2$ $18 - 6 = 12$
   $10 + 7 = 17$ $19 - 9 = 10$
   $18 - 6 = 12$ $17 - 4 = 13$
2. 8, 8, 8, 7
   18, 18, 18, 17
   6, 9, 6, 6
   16, 19, 16, 16
3. 4, 2, 1, 3
   14, 12, 11, 13
   4, 2, 5, 5
   14, 12, 15, 15
4. 8, 8, 9
   18, 15, 3, 18, 12, 7, 19
   2, 1, 5
   12, 16, 5, 11, 17, 2, 15
5. = < >
   < > =
   < = <
   > > =

Answers

Unit 3.5

1. Check that colouring matches calculations:
   9 + 3 = 12   8 + 6 = 14
   7 + 5 = 12   6 + 6 = 12
2. 15
   14
3. 1, 2, 12   2, 3, 13   4, 5, 15
   2, 2, 12   1, 8, 18   3, 6, 16
4. 13 13
   10 10
   13 13
   14 14
   10 1, 10
   14 4, 14
5. For example; 7 + 8 = 15
                 6 + 9 = 15

Unit 3.6

1. 9, 15   5, 13
   6, 15   5, 8, 13
   4, 7, 11   3, 10, 13
   7, 4, 11   10, 3, 13
   7 + 4 = 11   3 + 9 = 12
2. 13, 13, 12
   17, 11, 11
3. 11, 11, 12, 13
   12, 12, 13, 11
   13, 13, 14, 12
   14, 14, 15, 11
   15, 15, 16, 16
   16, 11, 17, 17
   12, 18, 11, 13
4. > < =
   > > =
   > > <
   = =
5. 10 15 14 12
   12 17 12 16

Unit 3.7

1. Check pupils' crossing out.
   8 8
   7 7
2. 7
   5
3. 2, 1, 9   1, 5, 5   3, 2, 8
   6, 2, 8   7, 2, 8   4, 2, 8
4. 9   3
   10 10
   9   3
   8   2
   10 10
   8   8, 2
5. 6   9
   10, 9   8, 9
   4, 11, 5   3, 15, 2

Unit 3.8

1. 13 − 5 = 8   15 − 7 = 8
   14 − 8 = 6   17 − 9 = 8
   13 − 4 = 9   12 − 3 = 9
   16 − 9 = 7   20 − 9 = 11
2. 6, 9, 8
   8, 3, 8
3. 9, 3, 7, 5
   9, 8, 8, 7
   9, 6, 8, 7
   9, 7, 4, 5
4. < < = >
   > > > >
5. 8 − 1 = 7   12 − 5 = 7   16 − 9 = 7
   9 − 2 = 7   13 − 6 = 7   17 − 10 = 7
   10 − 3 = 7 14 − 7 = 7   18 − 11 = 7
   11 − 4 = 7 15 − 8 = 7   19 − 12 = 7

Unit 3.9

1. 7 + 5 = 12   11 − 3 = 8
   13 − 4 = 9   4 + 8 = 12
   3 + 9 = 12   9 + 4 = 13
   20 − 8 = 12   12 − 5 = 7
2. 12 12   7   8
   15   8   9   8
   13   7   5 20
3. Greater   Less
   than 10:   than 10:
   5 + 7   15 − 6
   7 + 13   13 − 8
   9 + 2   18 − 14
   16 − 5
   8 + 4
4. 7   12   9
   13 15 17
   13   8   5
5. Answers will vary. For example:
   3 + 9 = 4 + 8 = 5 + 7

Unit 3.10

1. 16 − 7 = 9   12 − 9 = 3
   6 + 5 = 11   9 + 7 = 16
   14 − 6 = 8   7 + 5 = 12
2. (a) 6 + 7 = 13
   (b) 6 + 6 = 12
   (c) 12 − 3 = 9
   (d) 14 − 8 = 6
3. (a) 12 − 8 = 4
   (b) 9 + 6 = 15
4. Answers will vary. Check that they make sense.

Unit 3.11

1. 15, 6
   9 + 7 = 16, 16 − 9 = 7
   8 + 5 = 13, 13 − 8 = 5
2. 5 + 8 = 13   9 + 7 = 16
   13 − 5 = 8   16 − 9 = 7
3. 18, 2, 19, +9, −5, 19
   +1, −1, +8, −8, 12, −5, 12
4. 18 10, 14   9
   16 10   9, 18
   11   8   9
   4, 7 8, 16   9, 4, 13
5. (a) How many were left on the train?
       18 − 4 = 14
       14 + 4 = 18
   (b) How many (books) did she have altogether?
       15 + 4 = 19
       19 − 4 = 15
   (c) How many (cars) were left?
       14 − 8 = 6
       6 + 8 = 14
   (d) How many (strawberries) did she have left?
       13 − 8 = 5
       5 + 8 = 13

Unit 3.12

1. 6   9   16
   3, 3   4, 5   7, 9
   2, 1, 2   2, 2, 3   5, 2, 7
   16   20   12
   10, 6   12, 8   6, 6
   8, 2, 4   7, 5, 3   6, 0, 6
   18   18   19
   8, 10   9, 9   9, 10
   4, 4, 6   0, 9, 0   4, 5, 5
2. 19   20
   9, 10   11, 9
   4, 5, 5   6, 5, 4
   3, 1, 4, 1 2, 4, 1, 3
   20   20
   8, 12   9, 11
   3, 5, 7   4, 5, 6
   1, 2, 3, 4 3, 1, 4, 2
   For example: 17   19
                 7, 10   2, 17
                 3, 4, 6   1, 1, 16
                 1, 2, 2, 4 1, 0, 1, 15
3. Check pupil's own number walls.

Answers

Chapter 3 test

1. **(a)** 5, 10, 11, 13, 15, 17
   (b) 8, 12, 14
   13, 9, 7
   (c) 10, 9
   7, 15, 7
2. **(a)** 12, 17, 10, 8
   9, 18, 20, 10
   (b) 15   18
   8, 7   9, 9
   4, 4, 3   7, 2, 7
   (c) Lines should join:
   15 − 7 and 8
   17 − 9 and 8
   6 + 4 and 10
   8 + 8 and 16
   13 − 5 and 8
   14 − 4 and 10
   10 + 6 and 16
3. 7 + 6 = 13 4 + 8 = 12
   13 − 6 = 7 12 − 4 = 8
   16 − 7 = 9 2 + 11 = 13
4. 8, 9

Notes

Notes

Notes